THE CHINESE SKY DURING THE HAN

SINICA LEIDENSIA

VOLUME XXXVIII

THE CHINESE SKY DURING THE HAN

Constellating Stars and Society

BY

SUN XIAOCHUN AND JACOB KISTEMAKER

BRILL
LEIDEN · NEW YORK · KÖLN
1997

The authors are grateful for the financial support of the Society for Eurasiatic Celestial Sciences.

This book is printed on acid-free paper.

Library of Congress Cataloging-in-Publication Data

Xiaochun, Sun
The Chinese sky during the Han : constellating stars and society / Sun Xiaochun and Jacob Kistemaker.
p. cm. — (Sinica Leidensia, ISSN 0169-9563 : v. 38
Includes bibliographical references and index.
ISBN 9004107371 (alk. paper)
1. Astronomy. Chinese. 2. Constellations. 3. China—History—Han dynasty, 202 B.C.-220 A.D. I. Kistemaker, Jacob. II. Title. III. Series
OB17.X53 1997 96-40507
520'.931 dc21 CIP

Die Deutsche Bibliothek – CIP-Einheitsaufnahme

Sun, Xiaochun:
The Chinese sky during the Han : constellating stars and society / by Sun Xiaochun and Jacob Kistemaker. – Leiden ; New York ; Köln : Brill, 1997
(Sinica Leidensia ; Vol. 38)
ISBN 90-04-10737-1
NE: Kistemaker, Jacob ; GT

ISSN 0169-9563
ISBN 90 04 10737 1

PRINTED IN THE NETHERLANDS

"All people have a natural desire to know"

Aristotle (384-322 BC)
first sentence of *Meta Physica*

But ...

"Each person sees his own Truth,
speaking violates her,
testifying damages,
labeling might kill her,
believers bury her alive"

Lin Yutang

CONTENTS

CONTENTS

PREFACE

Constellations are the invention of human imagination. With constellations the sky is a unique document of ancient astronomy and ancient society. Therefore the study of constellations is important, from both a astronomical and a cultural point of view.

The Chinese sky of the Han is characterized by the large number of constellations, four times more than in the Greek sky of the same epoch. As a mirror of daily life it has much to tell us about the civilization and culture of the Han period. There have been many studies about Chinese constellations, from investigations of star maps and catalogues to research about the origin of the twenty eight lunar lodges. Most of them concerned the correlation of Chinese star names with Western ones, based on star catalogues from the Qing dynasty. Little effort has been made to understand the origin of constellations: their names, their astronomical and cultural context. This book addresses these issues.

In the first place, the authors aim to reconstruct the Chinese sky during the Han. To do this, they first studied the *Shi shi xingjing* star catalogue, the earliest in Chinese history. Their analysis confirms that this star catalogue was based on observations during the Former Han, rejecting the two-epoch conclusion held by some scholars, which in itself is a contribution to the study of the history of Chinese astronomy. Based on this *Shi shi xingjing* star catalogue, using descriptions of constellations in the *Tianguan shu*, *Jin shu*, *Kaiyuan zhanjing* and other sources, they reconstruct the sky during the Former Han, using various other historical star maps and catalogues to accomplish their gigantic task. Although it is impossible to identify each individual star from the available data, their reconstruction seems to give the first reliable representation of the Han sky. It is quite clear that it differs appreciably from reconstructions from the Qing catalogues. Thus we can see changes in Chinese constellations. From the point of view of positional astronomy and the study of ancient records of celestial phenomena, this reconstruction is useful and reliable.

Secondly, this book makes an attempt to reveal the cultural background of the Chinese sky. Careful textual research was done before the authors set out to do this, as we can see from the informative Appendices. The celestial projections show us the cultural background of terrestrial society. The Han period was the time of the formation of the complete Chinese sky. Many constellations, especially those of *Gan Shi*, were apparently constructed during that time. Therefore two contexts — astronomical (topographical) and cultural, are essential for the understanding of star names. This book is a first effort to consider these two contexts together, using the seasons as a connection between the sky and the terrestrial world. It is interesting to notice various features and structures in this sky and to know how they are related with philosophical ideas and social structures during the Han time.

Thirdly, as far as I understand, the authors want to continue this study to compare it with other Asiatic cultures. Indeed, the sky reflects very ancient myths and legends, with which the origin of constellations is certainly related. Probably we should understand the Chinese sky against the whole Asiatic background. For example, the origin of the twenty eight Lunar Lodges should be considered in its Asiatic cultural context. This sort of study will provide an answer to the problem whether the Chinese nomenclature of constellations grew up in isolation and independence or in communication with the outside world. This task will eventually need collaboration with Indologists, Assyriologists, Egyptologists, etc. But a good beginning can be made now from the Chinese sky, because the present examination conveys so much information useful for such a study.

The available historical material allows us to go as far back as the Han period. Therefore the Han sky should be the basis for any attempt to go back beyond the Han time. In this sense this book will provide a valuable reference for further studies. I am sure that the authors themselves will contribute much to such valuable cultural studies in the near future.

Keizo Hashimoto
Kansai University, Osaka
May 1996

FOREWORD

The construction of constellations has been a historical process. In the beginning they were seen as individual stars, eventually human imagination organized them into small groups (asterisms) which were always associated with various cultural aspects of human life and experience, customs and habits, religion and ideas, social structure, etc.. In the West these asterisms became the basis of the large constellations of later times. In China, however, asterisms in the same sense were called *xing guan*, Celestial Officials. They were so called because the ancient Chinese thought that *xing guan* had its specific function in the celestial world, just like a secular official who had his own duty in human society. The process of the construction of constellations was a long one. Asterisms grew in number and finally covered the whole visible sky. Constellations developed from asterisms, like a crystallization process.

The western constellations were consolidated in the classical Greek times. The Greek world was composed of many municipal states. There were alliances among them which frequently shifted. Unlike the Chinese rulers, however, their rulers did not think that they had much connection with the sky and the constellations. They left astronomers responsible for the demarcation and knowledge of constellations. Thus a unique system of constellations developed in the Greek world. Although there were different schools of astronomy, their definition of the constellations seems to agree.

It was different in China. At least from as early as the Xia, Yin and Zhou dynasties, Chinese rulers believed that they received a mandate from Heaven to rule over the world. About benevolent or disastrous rule, about strength or weakness, Heaven would warn them with Signs — omens near certain asterisms, such as solar and lunar eclipses, comets, meteorite showers, etc.. Astronomers (astrologers) had to explain to the rulers what was intended by Heaven. Because Chinese asterisms primarily came into being during the Warring States period, astronomers who served different masters built up rather different systems of asterisms. Therefore, from an academic point of view, different constellations represented different astronomical schools.

Due to many reasons, some systems of asterisms have been buried in oblivion. Some of them grew and prospered however. During the Han there were mainly three prominent systems, namely the systems of Shi (Shen) *Shi* (school), Gan (De) *Shi* and Wuxian *Shi*. There must have been systems of Huangdi *Shi*, *Hairen* (Marine People) *Shi*, etc., but because the historic material about these schools has been lost, it is nowadays impossible to investigate their systems of asterisms.

Since the Qin and Han dynasties, China became a unified and centrally governed empire. The imperial Han astronomers generally accepted Shi *Shi*'s system of constellations. The Shi *Shi* system consisted of most prominent constellations in the sky. Nevertheless, the other two schools, Gan *Shi* and Wuxian *Shi*, had many constellations which in an astrological sense were quite as important and which Shi *Shi* did not have. At the end of the Eastern Han, China was divided into the three kingdoms of Wei, Shu and Wu. Chen Zhuo, the royal astronomer of the Kingdom of Wu, made an effort to collect and conflate the constellations of the three schools. The result was a new large system of 283 constellations comprising 1464 stars. This system was adopted by all royal astronomical institutions of later times.

Because the large number of asterisms in Chen Zhuo's new system was not easy to remember and thus not easy to teach, some people made larger groups out of these asterisms. The most famous effort in that respect was the *Bu tian ge* (Song of Pacing the Heavens) from the Tang period. In this popular book all 283 asterisms were grouped into a so-called Three *Yuan* and Twenty Eight *Xiu* system, with each *Yuan* and each *Xiu* consisting of several asterisms, mainly for the purpose of education. But professional astronomers still used Chen Zhuo's system based on the asterisms of the Shi, Gan, and Wuxian schools.

The complete systems of asterisms of the Shi, Gan and Wu schools were established during the Han. If one wishes to know the history of astronomy in China, one should study the systems of asterisms. If one wishes to study the asterisms, one must start from the sky of the Han.

The present work is about the asterisms of the three schools during the

Han.

One of the authors, Sun Xiaochun, is a native of Liyang, Jiangsu, China. He graduated from the Astronomy Department of Nanjing University in 1984. Afterwards he studied with Professor Lu Yang for his Master's Degree in the History of Science. His thesis was *The meaning of Bei Dou in Chinese astronomy*. In 1989 he came to the Institute for History of Science of the Chinese Academy of Science in Beijing to study for his Doctor's Degree. He chose the Han sky as his subject.

In the meantime, by good luck, in Holland, ten thousand miles from China, the well-known physicist Professor Jacob Kistemaker, after retired from his position as director of the F.O.M. Institute for Atomic and Molecular Physics in 1982, returned to astronomy which he loved since he was young. He noticed that much of ancient Chinese astronomy had been neglected by the western world and decided to write a book on the Chinese sky. After this great effort was introduced in our institute by Dr. Wang Zhonglie (whose Doctor's Degree was supervised by Professor Kistemaker), it decided to help him with this effort. After some discussions we decided to introduce Sun Xiaochun to Professor Kistemaker. A Joint Education programme was arranged between China and Holland so that Sun Xiaochun was able to work in Amsterdam for one year. On the one hand Sun Xiaochun learned new methods, ideas and details of many cultures from Professor Kistemaker; on the other hand he used his own knowledge about ancient Chinese astronomy to work with Professor Kistemaker.

In the spring of 1992, Sun went to Holland. He worked very hard. He not only helped with the book, but also finished his thesis under the supervision of Professor Kistemaker. The doctoral ceremony was held in Beijing in the winter of 1993.

The two authors had collaborated fruitfully and decided to publish their work in two books. One has the title *The Chinese Sky During the Han*. The other has the title *The Origin of the Chinese Xiu* which will present more of Professor Kistemaker's initiatives and ideas. Now they decide to publish the first book. The other one will be prepared in due course and will be published in one or two years.

This book is the first specific study of the Chinese constellations written in English. Perhaps because of their different cultural backgrounds, western readers may be frightened away by the title of this book. But I want to remind them that both authors have been nurtured in modern astronomy. Professor Kistemaker is a scholar of highest repute in the west. He certainly knows the interest of western readers and how to cross the threshold to an exotic culture. Dr. Sun Xiaochun is a brilliant young scholar who is very quick to learn. Their collaboration is really like *a string of pearls and a girdle of jade* – a happy combination to bring out the best of each. Their book is a good guide for western readers who want to study the "mysterious" ancient Chinese astronomy. Lay readers who have more general interest in Chinese culture may find valuable material in it about social life, social structure, folklore and mythology which were all reflected in the sky. Perhaps this material can be found nowhere else or has never been discussed in this context. In short, the time and patience invested in this book will be compensated for with profit and pleasure.

People living in the East and in the West share one common sky above their heads. Constellations projected on this common sky sometimes differ very much due to several reasons: different cultures, different civilizations and immense difficulties in transportation and communication during thousands of years.

But all of this now seems to be in the past. The world is changing quickly and we are all living near to each other. It is about time to learn to know each other. Books like "*Chinese Sky during the Han*" give a unique opportunity to non-Chinese people to know more about our great history, and to Chinese people themselves to learn more about their ancient roots. In essence human beings are the same everywhere on this earth.

I believe this book will contribute to the Friendship between East and West. We appreciate the diligence of the Authors.

Bo Shuren
Institute for History of Science, Beijing
May, 1996

ACKNOWLEDGEMENTS

At the end of a long way to fulfilment my memories go back to several persons who were put on my Path of Life in a miraculous way. Two of them had a decisive influence. They were Wang Zhongli, a physicist of the Normal University of Beijing, who stayed in our FOM Institute for Atomic & Molecular Physics in Amsterdam between 1980 and 1982, and Yi Shitong, a historian-astronomer of the Ancient Observatory of Beijing who presented me his newly published Star Atlas in 1982 during my first visit to Beijing. Using his atlas I started my study on the Chinese sky. It took me nearly ten years of hard work, a few years together with Yang Zhengzhong of the CAS Astronomical Observatory in Beijing, to discover my basic lack of knowledge about the History of Chinese Astronomy.

Erik Zürcher, the director of the Sinological Institute of the Leiden University, was most helpful in those early days to explain the "Phenomenon China". I knew by experience that new discoveries and research happen on the boundaries between so-called specialisms. It needs some courage and diligence however to cross borders, and to listen to the criticism of annoyed colleagues not used to stupid questions. I am very thankful to Bob Kramers, Rik Schipper and Nathan Sivin who did not spare me but were quite helpful during those critical years.

In 1990, Wang Zhongli visited me again in Amsterdam. With his help I made up contact with Professor Bo Shuren, a specialist on the History of Chinese Astronomy of the CAS Institute for History of Science in Beijing. Our communication brought about a Joint Research Program between the Royal Netherlands Academy of Sciences (KNAW) and the Chinese Academy of Sciences (CAS) which started in January 1992. Then Bo Shuren sent his doctoral student Sun Xiaochun to Amsterdam as a guest researcher. (J.K.)

The present book is the result of our collaboration, which is a continuation of Bo Shuren's preliminary study on Sima Qian's *Tianguan shu* and ten years of hard work in Amsterdam to understand the Origin and Meaning of the Chinese Sky.

While this work is ready to be published we want to thank

— Professor Frans Saris, director of the FOM Institute for Atomic & Molecular Physics, and his coworkers, especially Tina Weeding and Jaap Sanders, who read the whole manuscript and polished the "English" text by "making it more American", Jan van Elst, Rob Lahaye and Bela Mulder for solving all computer problems for the final layout of this book, and Henk Sodenkamp and Herman Ficke for helping to make figures and illustrations.

— Professor Ed van den Heuvel, director of the Astromical Institute of the University of Amsterdam, and his coworkers Roelf Takens, Martin Heemskerk, and Tijn Smit.

— Professor Erik Zürcher of the Sinological Institute of the University of Leiden, for continuous stimulation and guidance. He provided us with several illustrations in this book.

— Professor Chen Meidong and Professor Liao Ke, former and present directors of the CAS Institute for moral support. And the colleagues Tian Jing, Jing Bing and Duan Yibing.

One of us (S.X.) especially wants to thank the KNAW and the Society for Eurasiatic Celestial Science for financial support.

We are grateful to Professor Bo Shuren for making another contribution to this book by writing the Foreword, and Professor Keizo Hashimoto for doing us the honour of writing the Preface. Last but not least, we thank Professor Nathan Sivin for his critical and valuable remarks which we used to improve our manuscript.

Jacob Kistemaker
FOM Institute for Atomic
& Molecular Physics
1009 DB POB 41883
Amsterdam, The Netherlands

Sun Xiaochun
CAS Institute for History
of Natural Sciences
137 Chao Nei Da Jie, 100010
Beijing, China

June 1996

LIST OF FIGURES

List of Figures

List of Tables

List of Illustrations

CHRONOLOGY OF CHINESE HISTORY

The Three Sovereigns, San Huang	Fuxi Nü Wa Shennong		
The Five Emperors, Wu Di	Huangdi Zhuanxu Ku Yao Shun	27th century BC 24th century BC	
	Founded by		*Capital*
Xia dynasty	Yu	21st century BC	
Shang (Yin) dynasty	Tang Wang	16th century BC	Anyang
Western Zhou dynasty	Wu Wang	11th century BC	Changan
Eastern Zhou dynasty		770-221 BC	Luoyang
Chunqiu, Spring & Autumn Period		770-476 BC	
Zhanguo, Warring States		476-221 BC	
Qin dynasty	Qin Shi Huangdi	221-207 BC	Xianyang
Han dynasty			
Western Han	Liu Bang	206 BC - 8 AD	Changan
Xin dynasty	Wang Mang	9 AD - 25 AD	Changan
Eastern Han		25-220	Luoyang
Sanguo (three kingdoms)		220-280	
Jin dynasty		265-420	Changan
Southern & Northern dynasties		420-581	
Sui dynasty		581-618	Changan
Tang dynasty		618-907	Changan
Five dynasties		907-960	
Song dynasty			
Northern Song		960-1127	Kaifeng
Jin		1127-1211	
Mongol invasion (Kublai Khan)		1211-1280	
Southern Song		1127-1279	Hangzhou
Yuan dynasty (Mongol)		1280-1368	Beijing
Ming dynasty		1368-1402 1402-1644	Nanjing Beijing
Qing dynasty (Manchu)		1644-1911	Beijing
Republic		1912-1949	Nanjing
People's Republic		1949-present	Beijing

CHAPTER ONE

INTRODUCTION

1.1 What is the Chinese sky?

The Chinese, like other people, have lived in a particular domain on earth. But the sky is not specific to them. It is the same to all people living at the same latitude. With the naked eye people can see roughly 3000 stars in the sky. What made the sky *Chinese* was the way they looked at it, the way they configured constellations from the stars and named them, the way they organized these constellations and the way they interpreted celestial events which happened against their self-conceived celestial background.

Chinese star names differ widely from the occidental ones. Instead of mythological heroes and superhuman beings, you find objects from the realistic terrestrial world: people of all classes, as high as the emperor, the royal family, officials and generals, as low as peasants and petty soldiers. All kinds of social institutions, constructions, activities and objects, either for daily use or ceremonial occasions, were installed. One should not be astonished to find the Celestial Toilet and the Pit of Manure in the holy sky because it was meant to be a complete projection of the terrestrial world.

These remarkable facts have attracted the attention of curious people asking "why?" It happens that we are among those curious people and have set out to make an effort to understand the situation. We will treat the Chinese sky as a *phenomenon;* as in doing physics we try to explain this phenomenon, using basic material, analyzing, hypothesizing and eventually approaching the truth.

Firstly, the *phenomenon* of the ancient Chinese sky is a *natural* one. The 283 constellations are spread out all over the visible sky. Not only the star names but also their positions and celestial surroundings should be considered in order to explain the meaning of these constellations. Some constellations were simply descriptions of the celestial sphere. For instance *Beiji* 北極, the North Pole Office indicated the position of the pole. The most important fact is the correspondence between constellations and seasons. As the sun moves among the stars along the ecliptic, the part of the sky which is visible at night shifts with the change of the seasons. Thus the constellations in that part of the sky presumably bear the touch of the corresponding season: the climate, natural life and human activities. The correspondence between the sky and the seasons is the primary key to understanding star names.

Secondly, the Chinese sky is a *historical phenomenon.* The constellations were designed neither during a single short period, nor by a single astronomer or astrologer. They developed during a long period of history. Some constellations like *Beidou* 北斗, *Shen* 参 and the four cardinal asterisms, evidently belong to the oldest ones in the sky. The four cardinal asterisms are probably from the 24th century BC. During the Zhou 周 period (1027 - 221 BC) astronomy developed considerably. The gnomon and the water clock were used in astronomical observations; the twenty eight lunar lodges (*xiu* 宿) were introduced as celestial reference points to register the positions of celestial objects. All astronomical measurements seemed to serve primarily the purpose of calendar-making and astrology based on the movement of the planets. During the later Zhou, namely the Zhanguo 戰國 (Warring States) period (475 - 221 BC), the sovereignty of the Zhou ceased to exist except in name. Each state had its own court astrologer, among them were Shi Shen 石申 of the Wei 魏 state and Gan De 甘德 of the Qi 齊 state. Apparently such a situation provided a nutritious soil in which all kinds of astronomical and astrological ideas could germinate. About 200 constellations in the Chinese sky were attributed to Shi Shen and Gan De. Although we will prove that this attribution is not valid, it does indicate that at that time more and more constellations were configured. The real development of the Chinese sky happened during the Han, however. Historical aspects should be taken into account to understand the names and configurations.

Finally, in a broader sense, the Chinese sky is a *cultural phenomenon.* It reflects a cultural complex, a celestial representation of the terrestrial society. The sky presents an astrological scheme of constellations which was

based on the principles of cosmology. The universe was a harmonious unity of *Tian* 天 (Heaven), *Di* 地 (Earth) and *Ren* 人 (Man). This unity was the basis for the mutual response between Heaven and Man. Chinese astrology was strictly omen-astrology which only dealt with affairs of state. The first step was to observe portents in the form of celestial or meteorological phenomena; the second step was to correlate these portents with events in the imperial society. Since the observed phenomena happened on the celestial background, astrologers designed the sky as a counterpart of the terrestrial society. The sky was like the setting of a stage on which all kinds of events were happening. Thus the astrological interpretation of these events became directly related with affairs of state.

It would be very interesting to know how astrologers "designed" the sky by configuring and naming constellations. The construction of constellations was based on the understanding of nature, as we have said, primarily based on the understanding of the seasons. Some old constellations reflected ancient myths. Mythology was nothing else than a way to describe and explain the influence of nature on human society. It was the early stage of the human mind.

Later on, philosophy and cosmological ideas were incorporated into the construction of the sky. A typical Chinese way of explaining nature was with the *yin-yang* 陰陽 (Feminine and Masculine Forces) and *wu xing* 五行 (Five Elements) theories. Astrology also made use of these theories; the interpretation of *omina* mainly derived from cosmological and philosophical principles.

1.2 Why choose the Han period?

The choice was based on historical considerations. The Han time was the most important period for the formation of *Chinese* culture. The social structure, the cosmological ideas, philosophies, rituals and ceremonies, etc., were cast at large; changes and modifications were all based on Han models. The celestial representations of these Han models changed very little indeed. To understand the Chinese culture from its roots, one should go back to the Han time.

Before the Han, the period of philosophers yielded many political, social and philosophical ideas. There was already a tremendous amount of knowledge in separate branches of learning such as medicine and astronomy. It is still a question how this knowledge was acquired. Presumably cultural exchanges with other civilizations happened, but little is known about this episode. It was during the Han that the academic tendency to systematize all kinds of knowledge by means of cosmological principles started. The *yin-yang* philosophy and the Five Element theories became the fundamental principles used to explain all kind of phenomena, natural as well as social. Everything could be explained by the interplay of the *yin* 陰 and *yang* 陽 forces and the Five Elements. An obvious example is medicine. The traditional medical classic *Huangdi neijing* 黄帝内經 was totally based on the *yin-yang* and Five Element philosophies. But certainly the knowledge of medicine was not derived from the *yin-yang* philosophy but from long experience. The book was only an attempt to systematize this knowledge.

Original Confucianism dealt mainly with political and ethical problems. But an important development of Confucianism happened by the incorporation of the *yin-yang* and Five Element theories. Thus Confucianism also possessed the "scientific" tools to "investigate the *boundaries* between Heaven and Man."[1] Astronomy, astrology and calendrical sciences had always been "official", exclusively controlled by the court. Now the new elements of Confucianism enabled it to investigate the heavens as well. This contributed to its establishment as state ideology during the Han. Consequently, astronomy and astrology developed dramatically.

Moreover, the Han period was a time of confluence of knowledge from different sources and different schools. The unification of a vast land with many different cultures formed the basis for the so-called Han traditions[2]. During that period there was a prince, Liu An 劉安, who organized hundreds of scholars from different places to write a book *Huainan zi* 淮南子 (Book of the Prince of Huainan) which contained comprehensive knowledge. There was also an astronomer Luoxia Hong 落下閎, the inventor of the armillary sphere, who was summoned to court from the remote south-

[1]Sima Qian 司馬遷, *Shi ji* 史記 (Historical Records), ch.130.

[2]Before the Han was the unified Qin empire. Of course the measures taken by Qin Shi Huang Di 秦始皇帝 had a great impact on the formation of Chinese culture. But the empire only lasted 15 years. The real accomplishment in this sense was achieved during the Han. In fact, historians always consider these two empires together calling it the Qin-Han period.

Figure 1.1: Sima Qian 司馬遷 (145 - 87 BC) writing the *Shi ji* 史記 (Historical Records). In doing so he desired to "investigate the *boundaries* between Heaven and Man, to thoroughly understand social changes ancient and modern, and finally to get a total perspective (of the Universe) of my own." This picture is reproduced from a postage stamp issued by China Post in 1994. The original words of Sima Qian are also shown on the stamp.

western region Ba 巴 by Emperor Wu 武 for the purpose of calendar reform. These examples demonstrate that it is very likely that knowledge from other civilizations might have diffused in. This situation makes the study of the Han time so interesting.

The choice of the Han period also emerged from the general opinion that the complete system of the Chinese sky was formed during that period, which we will explain in due course. Although during the preceding Zhan Guo period there must already have been quite a lot of constellations, their names remained quite obscure except for the *xiu* 宿 (Lunar Lodges or Lunar Mansions). The first systematic description of the sky was by Sima Qian, the *Tianguan shu* 天官書 (Monograph on Celestial Officers (constellations)). Many constellations which have been attributed to Shi Shen and to Gan De were apparently developed after the *Tianguan shu*. The 283 constellations were from three astronomical schools which were active during the Han but attributed their knowledge to Shi Shen, Gan De, and Wuxian 巫咸, the last one a vague personality from the Yin 殷 dynasty, about 1200 BC. Immediately after the Han, the astronomer Chen Zhuo 陳卓 constructed the constellations of the whole sky based on these three schools. From then on the Chinese sky was fixed.

The Han sky is retrievable because in the first place there exists an ancient star catalogue - the *Shi shi xingjing* 石氏星經 star catalogue - which gave the coordinates of the leading stars of the constellations of Shi Shen's school and of the 28 *xiu* 宿. This catalogue proved to be based on observations during the Han. Its constellations generally included bright stars. Historical star maps and star catalogues from following dynasties facilitated the reconstruction of the Han sky.

1.3 Why study the Chinese sky?

The interest is twofold: from a point of view of pure positional astronomy and from a point of view of historical studies. The latter concerns the origin and the meaning of Chinese star-names and their historical, cultural and mythological backgrounds.

1.3.1 Positional astronomy

From an astronomical point of view, ancient records of astronomical observations sometimes are very useful for the purpose of modern research. The great virtue of ancient Chinese astronomy has been the assiduous observation and well preserved records of all kinds of celestial phenomena, such as eclipses, comets, conjunctions of planets, novae and supernovae, etc. These records provide invaluable material for modern astrophysical studies. One of the parameters of a celestial event is its position on the sky. Most celestial phenomena were recorded with constellations as reference background. For instance, the well-known "guest star" (nova) of the year 1054 was said to be visible near the constellation *Tianguan* 天關 (ζ Tauri). The immediate question which arises is which star was *Tianguan* at that time.

The Chinese sky has been changing during history in the sense that some star names could have denoted different stars. The number of stars and shapes of some constellations were also changing. One should be aware of this when making use of ancient records of celestial phenomena. Several researchers have made efforts to identify the constellations; the most recent one was Yi Shitong 伊世同[1981] who provided a correlative star catalogue and star atlas between China and the West. This excellent work proved to be a very useful reference for the study of the Chinese sky. But his identification work was based on the *Yixiang Kaocheng* 儀象考成 (General

Description of Imperial Instruments) from the Qing dynasty which does not necessarily correspond with the sky from before the Qing.

Here is an example to explain our argument. The star *Langjiang* 郎將 on the map of Yi Shitong is identical with 31 Coma Berenices, a star of 5th magnitude. The *Tianguan shu* described *Lang Jiang* as a "great star" which meant it was among the brightest stars in the sky. According to our study, *Langjiang* was identical with α Canes Venatici, a star of the 2nd magnitude during the Han. One could consider it as a supernova if one were unaware of the error in celestial position. Thus from a purely astronomical point of view the study of the Han sky is necessary. Indeed, for a reliable use of ancient records of celestial phenomena in China the study of the sky from all periods in history is desirable.

1.3.2 Historical study on star-names

The Chinese sky, with such a large number of constellations named after terrestrial objects, is a unique document for a historical study. A variety of aspects of ancient civilization and culture are reflected in this total picture of the sky, such as myths and legends, daily life and natural knowledge, rituals and ceremonies and state institutions and officialdom. It is certainly of interest to compare these terrestrial backgrounds with the celestial projection, and to explain and understand it.

One question has already appealed to the interest of scholars: Did the Chinese star-name system develop independently or through influences from other cultures, possibly from Babylon, Iran or India? By comparing similarities between Chinese and European (Greek) nomenclatures, scholars have tended to the conclusion that the Chinese nomenclature of the constellations represents a system which grew up in comparative isolation and independence. Schlegel [1875] has even gone further to prove that the Chinese system was the origin of all systems. Most of his arguments, however, seems to be far-fetched. He selected what seemed to suit his postulate from miscellaneous astrological descriptions.

Babylonian astronomy (or astrology) matured much earlier than Chinese astronomy. A transmission of part of Babylonian astrology to China before the Han time can not be excluded. The striking similarities between Babylonian and Chinese astrology, namely the empirical approach, the "official" character and the hereditary position of astrologers, suggest an early interchange of ideas between the Near East and the Far East. Nevertheless, a conclusion can only be drawn after carefully studying the Chinese sky and comparing with the ancient Babylonian sky[3]. We do not intend to treat this problem in depth in this book, but the question itself shows the manifold links of the Chinese sky with ancient civilizations, not only Chinese, but also Asiatic.

1.4 What has been done before on the subject?

Earlier studies concerning the identification of Chinese constellations and the analysis of the *Shi shi xingjing* star catalogue will be discussed in Chapter 3. Here we only outline those studies which were aimed at understanding Chinese star names.

As early as 1875, Gustav Schlegel made a tremendous effort to study the star names. In his huge volume *Uranographie Chinoise* he made a complete survey of Chinese star names which had appeared in ancient literature. About 760 star names were identified based on textual research and about 700 other variant star names were mentioned. He managed to correlate all these star names with western ones basically according to the star map in the *Tianyuan lili quanshu* 天元曆理全書 (Complete Treatise on Calendars) by Xu Fa 徐發 (early Qing). The number of star names became very large because he counted all synonyms and variant astrological names. These names, of course, sometimes helped to understand the meaning of the constellations, but in most cases only added confusion.

Schlegel's major effort was to search for as many analogies as possible between the Chinese and the European nomenclatures, the latter, as he correctly pointed out, actually being derived from ancient Egyptian and Babylonian nomenclatures. Thus his book was the first comparative study of constellations between China and the Middle East. Some analogies he found are inspiring. For instance, the *xiu Bi* 畢 (ϵ Tau; Hyades) was worshipped as *Yu Shi* 雨師 (God of Rain) in China, which seemed to be identical with the god of rain in Greece or Rome[4]. Still this divinity was not

[3]A reconstruction of the Babylonian sky of the 24th century BC was made by Papke [1989] based on his study of the Mul · Apin clay tablets.

[4]Schlegel, G., *Uranographie Chinoise*, Brill, Leyden, 1875, p.368.

natively Chinese or Greek, but derived from western Asia where the early rains coincided with the heliacal rising of these stars[5]. Schlegel's work has been widely referred to by researchers.

Schlegel's approach was to put the Chinese constellations in the climatological and social context of ancient times and to compare them with lores and myths from various Eurasiatic cultures. His approach may be called "*astro-mythological*". But his chronology seems impossible. He estimated that the Chinese *xiu* names originated from 16000 years ago which is highly improbable.

After Schlegel's work came the French translation of the *Tianguan shu* by E. Chavannes in 1898. This translation is important because he introduced the Chinese constellations with their entire astrological context for the first time into the western world. The *Tian Guan Shu* was the earliest systematic astrological description of the Chinese sky. Chavannes' work was by no means a simple translation; it incorporated a lot of enlightening information.

The most recent work towards this end was by Ho Peng Yoke [1966]. He translated the astronomical chapter of the *Jin shu* 晉書 (History of the Jin Dynasty) into English with valuable comments in footnotes. This chapter was originally written by Li Chunfeng 李淳風 from the Tang dynasty. It was a complete astrological description of the Chinese sky including all constellations of the three schools from the Han. Thus it is an invaluable reference for our present study. Ho Peng Yoke's study of the astronomical chapter of *Jin Shu* is of comparable importance to Chavannes' study on the *Tianguan shu*. The translation of star names is the first step to understand the Chinese sky.

1.5 Purpose and method of the present study

As one may anticipate from the preceding description, the present study has two purposes. One concerns positional astronomy: What did the Chinese sky look like during the Han? The other, concerns its cultural aspects: what was the meaning of the Chinese sky and what was the philosophy behind it?

[5]R.H. Allen, *Star Names: Their Lore and Meaning*, Dover Publication Inc., New York, 1963 (republication of the 1899 edition), p.389.

The basic approach of our present research is historical. Like any other historical research, textual research is fundamental. All material should be weighed and analysed. But one should not be addicted to it. Since so much source material has been lost during history, it would be silly to refuse any assumption simply because it was not found in the existing ancient literature.

There are two features of ancient astronomy which determined the methods with which we study the ancient sky: the exactness of the astronomical data and the interrelation with other aspects of ancient culture. Our method of study can be generally outlined as follows:

1. *Astronomical analysis*

Astronomical data can be used for the determination of the time of the observations. For the constellations of Shi Shen's school there exists a star catalogue entitled *Shi shi xingjing*. The analysis of the catalogue is crucial for the study of the Han sky.

Some star names can be understood simply by their position in the sky. A constellation near the celestial pole was called *Beiji* 北極 the North Pole Office, for example. Such star names were descriptions of the sky by themselves. Descriptions of the relative positions of stars should be considered against the celestial background of ancient times.

2. *Seasonal considerations*

The seasons regulate man's life. Sowing and harvesting must be done in the right seasons. Rituals and ceremonies happened at fixed seasons because that was the way to follow the Way of Heaven. Since the sky was related with the seasons, the picture of social life could be projected onto the sky according to the seasons.

There were essentially two ways of connecting the seasons with the sky: the *Solar Calendar* and the *Lunar Calendar*. By the Solar Calendar we mean that climatological and social happenings are connected with the sky as a function of the position of the sun on the ecliptic. The Lunar Calendar does the same according to the position of the anti-sun, where the full moon appears. The two calendars are half a year in season and 180 degrees in position out of phase.

Precession causes the cardinal points to shift westward 30 degrees dur-

ing approximately 2000 years. This corresponds directly with a change of one month in the climatological meaning of the constellations in the zodiacal belt.

3. *"Astro-mythological" explanations*

One cannot understand the meaning of the Chinese sky without knowing its astronomical, astrological, philosophical and mythological background. Literal translations of the star names can only be considered as indications of certain images. Explanation is necessary and it should be based on considerations of the neighbouring constellations, as well as on those on the opposite side of the sky[6]. Quite often constellations on opposite sides of the sky told the same story because both the solar calendar and the lunar calendar were applied.

The Chinese sky has been developed and modified by Han astronomical schools to fit into the Chinese tradition, history and way of thinking. But originally, in its primitive mythological sense, it reflects the lore and tales of the Eurasian people. For a real understanding of the original meaning of the names of the older constellations of Shi Shen's school and the 28 *xiu*, a comparison with classical Chinese, Indian and Middle Eastern mythology will be necessary. The influence of the Chinese Five *Di* 帝 mythology as well as some aspects of solar- lunar mythology are also recognizable in the sky.

The Han additions visible as constellations of Gan De's and Wuxian's astronomical schools reflect the tendency to revise the ancient mythology in a Confucian way by adopting the *yin-yang* and Five Element philosophies. Most of this should have produced a celestial scenery which represented the idealized Zhou 周 Empire and its Court.

4. *Analogous study of the sky of other Asiatic civilizations*

Ancient civilizations were not isolated as some researchers seem to think. The existence of analogous lunar lodge systems, probably all originating from the last millennium BC, indicates a lively communication between these widespread peoples. The pre-Han Chinese sky, visible in the con-

[6]Leopold de Saussure has pointed out that Chinese *xiu* are coupled with each other on opposite sides of the sky. This certainly strengthens our argument to consider constellations on opposite sides of the sky together. See L. de Saussure; *Les Origines de l'Astronomie Chinoise*, Paris, 1925.

stellations of Shi Shen's school reflects the same stories as one can find in the Hindu-Vedic or in the Sumerian sky. We are not ready to draw conclusions based on these analogies in this book, but we want to emphasize that it is sometimes helpful to know them when trying to explain shapes and positions of constellations in the Han sky and to understand the mysterious names of some *xiu* constellations.

We want to point out that the present book only represents the first stage of our study. Generally speaking this book is a historical study. We use historical material including star catalogues, star maps and astrological texts to trace back to the Han time in order to get a general picture of the Han sky. This is indispensible for the second stage of our study - to have an "astro-mythological" explanation and understanding of the Chinese sky against the climatological and social background of Asiatic cultures. Of course the two stages of our study are not isolated from each other: a thorough historical research of the Chinese sky will certainly provide a sound basis for "astro-mythological" study and the total picture of ancient Asiatic cultures sometimes helps to provide a proper understanding of the Chinese sky. Nevertheless, in this book we intentionally limited our study to the Chinese historical background.

1.6 The structure of the book

The structure of the book resembles the process of our present study. Chapter 2 is a brief history of the Chinese constellations. It is intended to be an introduction to readers who encounter the Chinese sky for the first time. It is also intended to explain the importance of the Han period for the formation of the Chinese system of constellations. Furthermore, by mentioning and briefly introducing all valuable and important historical star maps and catalogues we demonstrate that the reconstruction of the Han sky is possible and practicable.

Chapter 3 concentrates on the study of the *Shi shi xingjing* 石氏星經 star catalogue. Our new analysis of the catalogue by means of Fourier transform has settled the dispute over the observation time of the catalogue. The observation was made during the Han. The leading stars of the constellations of *Shi Shi* 石氏 will be the footholds for the reconstruction of the constellations. These constellations were actually the main constellations in the Han sky.

In chapter 4 we continue to study the constellations of *Gan Shi* 甘氏 and *Wuxian Shi* 巫咸氏. They were extensions and ideological filling-ins of the framework of *Shi Shi*. By careful study of descriptions of their relative positions and by using available star maps and star catalogues from history, we reconstructed the constellations of *Gan Shi* and *Wuxian Shi*. The constellations of the three schools are presented in the star map for an epoch of 100 BC, attached to the end of this book. This is our reconstruction of the Han sky. The identification of the Chinese stars with modern star names is clearly visible on the map.

Based on this star map we began to investigate the cultural aspects of the Chinese sky. We aimed to get a better understanding of the star names based on the Chinese cosmological, mythological and historical background rather than a simple translation of star names. Our results also support the general conclusion that the complete system was formed during the Han; the Han ideology is obviously recognizable in the structure of the sky. Chapter 5 discusses the general philosophy behind the Chinese sky and chapter 6 considers in detail its structures and groups of constellations which have mythological and social meanings.

It is impossible to present details of the historical research concerning the constellations in the main text. So we present them in three appendices for each school. Translations of star names, historical evidence and necessary explanations according to our study are given.

1.7 Technical remarks

In this book all Chinese words and characters are romanized according to the standard *pinyin* 拚音 (phonetizing) rules widely and officially used in China nowadays, except in some references where the Chinese characters were originally romanized according to other rules. However, as a convention in historical studies, the traditional Chinese characters, not the simplified ones, will be used.

A separate index has been created for Chinese constellation names which gives their corresponding locations among the western constellations and their sequential numbers as listed in the appendices. Thus in the general index most Chinese star names are not included.

For reference to the "Standard Histories" of ancient dynasties, such as *Shi ji*, we primarily used the modern edition of the Zhonghua Shuju Press in Beijing.

We want to emphasize that our work is primarily based on available Chinese sources, which is necessary because of the nature of our subject. Discussions can be extended by comparing with important publications from authors like Chr. Cullen, W. Hartner, J. Needham, N. Sivin, F. R. Stephenson and M. Teboul, but that is not our meaning. We try to look into original sources as much as possible within the limited time. Thus very probably we have missed some recent valuable publications on the relevant subjects.

Bronze celestial globe from the Kangxi period (1673) of the Qing dynasty. It was designed by Nan Huairen (Ferdinand Verbiest, 1623-1688, a Belgian jesuit), now exhibited at the Ancient Observatory in Beijing. Both the traditional Chinese constellations and the newly introduced constellations of the south-polar region were inlaid on this globe.
Photo taken from the *Zhongguo gudai tianwen wenwu tuji* (Wenwu Press, Beijing; 1980).

CHAPTER TWO

BRIEF HISTORY OF CHINESE CONSTELLATIONS

2.1 Stars from remote antiquity

It is commonly known that stars in the sky indicate seasons. Certain stars have been observed by people in China to determine the seasons since remote antiquity. Archeological evidences of observations of stars are available on pottery from the Neolithic ages on which certain patterns of groups of stars were painted and from oracle-bone inscriptions from the Yin time (1300 - 1208 BC) in which certain names of stars were mentioned[1]. The oldest records of star names were found in one of the ancient Chinese classics, the *Shangshu yaodian* 尚書堯典 (Canon of Yao in the Book of Documents). They are the names of the four cardinal asterisms which have already been studied by many researchers since the eighteenth century[2]. These stars are related to seasons, as the text literally goes:

> The day is of medium length and the star is in *Niao* 鳥 (Bird). You may thus exactly determine mid-spring. ... The day is at its longest and the star is *Huo* 火 (Fire). You may thus exactly determine mid-summer. ... The night is of medium length and the star is in *Xu* 虛 (Void). You may thus exactly determine mid-autumn. ... The day is at its shortest and the star is in *Mao* 昴 (Hair). You may thus exactly determine mid-winter[3].

[1]Pan Nai 潘鼐, *Zhongguo hengxing guance shi* 中國恆星觀測史 (History of Stellar Observation in China; Shanghai, 1989), pp. 1-3.

[2]To mention a few of them: A. Gaubil [1732], J.B. Biot [1862], Shinjō Shinzō 新城新藏 [1933], Zhu Kezhen 竺可楨 [1926], and Iijima Tadao 飯島忠夫 [1928].

[3]The translation is taken from James Legge's [1865] translation of the *Shu jing* 書經 (Book of Documents) in the modernized edition by C. Waltham; *Shu Jing - Book of History*,

The stars *Niao* 鳥 (Bird)[4] *Huo* 火 (Fire), *Xu* 虛 (Void) and *Mao* 昴 (Hair) are commonly recognized as the four cardinal asterisms of the 28 Chinese lunar lodges (*xiu* 宿); they were identified with the *Xing* 星 *xiu* (α Hya) for *Niao*, the *Xin* 心 *xiu* (α Sco) for *Huo*, the *Xu* 虛 *xiu* (β Aqr) for *Xu*, and the *Mao* 昴 *xiu* (Pleiades) for *Mao*. The text has been considered by most scholars as a set of real records of the culmination of these stars at dusk for respective seasons. Thus by means of precessional calculation scholars have come to different assertions about the date because they assumed different times after sunset. We mention a few examples:

— A. Gaubil[5] dated these observations as between 2155 and 2796 BC, which includes the age of the legendary Yao Di 堯帝 (c. 2300 BC);

— Zhu Kezhen[6] determined the date for the first three stars as the beginning of the Zhou 周 dynasty (c. 1000 BC), but suggested *Mao* could have originated from the time of Yao Di;

— Iijima Tadao 飯島忠夫 [7]. concluded 400 BC, suggesting influence from the Middle East

In our opinion the text should not be considered as a set of records of the culmination of these stars. Such meridian observations by means of a gnomon (*biao* 表) started in the Middle East as well as in China in the 7th or, at the earliest, 8th century BC. Thus apparently the text in the *Yaodian* had already

---

Henry Regnery Co., Chicago, 1971, pp.3-7. The only exception is that we read "the star is *Huo*" rather than "the star is in *Huo*" because the Fire star (*Huo*) was no doubt α Sco.

[4]By *Niao* the Bird was probably meant the southern palace *zhu que* 朱雀 (Red Bird) including seven *xiu* constellations; but historians all agree to identify it with α Hya of the *Xing* 星 *xiu* since the star is located in the middle of the palace.

[5]A. Gaubil, *Traité de l'astronomie Chinoise;* Paris, 1732.

[6]Zhu Kezhen 竺可楨, "Lun yi shuicha ding Shangshu yaodian sizhongxing niandai" 論以歲差定尚書堯典四仲星年代 (On the date of the four cardinal asterisms in the Canon of Yao), *Ke Xue* 科學 (Science), 11 (1926), No.12, pp. 100-106.

[7]Iijima Tadao 飯島忠夫 was of the opinion that Greek Babylonian influence had a big impact on the formation of Chinese astronomy about the time of Alexander the Great (about 300 BC). See Iijima Tadao, "Shina no jōdai ni okeru Girisha bunka no eikyō to jukyō keiten no kansei" 支那の上代における希臘文化の影響と儒教經典の完成 (Greek influence on the ancient civilization of China and the compilation of the Confucian classics with reference to calendrical science and astronomy), *Tōyō Gakuhō* 東洋學報 (Reports of Oriental Society of Tokyo), 11 (1921), No.1, pp.183ff.

Table 2.1: Position of the cardinal points in the *Yaodian* 堯典 (Canon of Yao).

Lunar Lodge (*xiu*)	Star	Right Ascension	Deviation from
or Name of Star		at 2400 BC	the cardinal points
Mao	η Tauri	23h45m	-15m
Niao (Xing)	α Hydrae	5h42m	-18m
Huo (Xin)	α Scorpionis	12h18m	+18m
Xu	α,β Aquarii	17h47m	-13m

been reorganized by later compilers who had current astronomical knowledge. But the text contained the hard facts: the four cardinal asterisms[8]. Thus for age determination of these asterisms, there is no confusion such as the assumption of the time of dusk. We calculated the Right Ascensions of the four stars. At the year of 2400 BC the values fit the cardinal points rather well, as is shown in Table 2.1. The mean deviation of these four cardinal points is -7 minutes, which corresponds with 140 years of the precessional shift. It suggests a first use of these cardinal asterisms about 2300 BC, with an uncertainty of plus or minus 250 years. This includes the reign of Sargon

[8]We believe that the four cardinal asterisms must have been noticed and used by peoples all over Asia in the second and third millennium BC. Actually they were constellations on the primitive Lunar and Solar calendars of nomads and first farmers in the 3rd millennium BC.

When the full moon passed by the Pleiades (*Mao* 昴) it was the mating time for the Ram and the Ewe, to be followed by an autumn festival where the Ram was offered to the gods. This event was biologically determined within two or three weeks between the middle of September and the second week of October to get lambs about the first of March. About the time of Yao Di in China, as well as of Sargon in ancient Akkadia (Sumeria) (2300 BC) these weeks about the autumn equinox usually had a full moon at the Pleiades (*Mao*). Therefore the Pleiades were a landmark on the calendar of pastoral nomads of the 3rd millennium BC, which we note as a *cardinal point*.

The same holds, on the opposite position on the sky, for the red star Antares (*Huo* 火) which even became the base of an early pastoral-nomadic calendar which prescribed refreshment of the family fire-place (to worship the god of fire). See Pang Pu [1990].

The other two cardinal points on the sky corresponded with the position of the Full Moon or the Sun on respectively Mid-Summer and Mid-Winter, high days for bird catching and deer hunting. We will come to details of this in our second book.

in Akkadia, as well as of Yao Di[9].

Some more star names could be found in the *Xia xiao zheng* 夏小正 (Small Calendar of the Xia Dynasty) which was generally regarded as including calendrical material from the Xia. The culmination at dusk and dawn, and the rising and setting of some conspicuous stars at sunrise or sunset, were recorded in the order of the twelve months of the year. Several more star names, such as *Shen* 參 *xiu* (Orion), *Beidou* 北斗 (Ursa Major), *Zhinü* 織女 (Vega) and *Nanmen* 南門 (α Cen), were mentioned in the text[10].

Star names could also be found in other books from before the Han, such as *Zuo zhuan* 左傳 (Master Zuo's Complement of the Spring and Autumn Annals), *Guo yu* 國語 (Discourses on the States) and *Shi jing* 詩經 (Book of Odes). In the *Shi Jing* there are sentences like "In the seventh month the Fire star (*Huo* 火) is descending westward, and in the ninth month it is the time to wear more clothes (*qi yue liu huo; jiu yue shou yi* 七月流火; 九月授衣)."[11] Actually the Fire star served as the season-telling star for the Yin people and the tradition was maintained by the descendants from the Yin dynasty[12].

As far as the existent ancient literature is concerned, all star names mentioned were from the 28 *xiu*, except for a few extraordinarily bright stars like *Beidou* and *Zhinü*. The purpose of observing these stars was obviously to determine the seasons or to make a calendar. It seems that there was no intent to observe or to register stars in the total sky before the Han or, safely speaking, before the Zhanguo (Warring States) period (480 - 222 BC). The construction and naming of constellations in the whole sky has apparently been an attempt by astronomers and astrologers after the Zhanguo period,

---

[9]It is interesting to notice that this date also corresponds with the ancient Sumerian Sky. The Mul · Apin clay tablets described the Sumerian sky at about 2300 BC. See Papke [1989] and Van der Waerden [1974]. Also the Indian system of *Naksatra*, lunar mansions, might refer to that time. Needham suggested that these cardinal stars "might possibly be a part of the traditional patrimony of knowledge about the heavens derived from Babylonian sources." See *Science and Civilization in China*, vol.3, section 20, p.177.

[10]See Noda Chūryō 能田忠亮 [1941]. Moreover Chen Zungui 陳遵嬀 [1984] gives the opinions of other researchers on this subject.

[11]*Shi jing* 詩經: *Guo feng* 國風 (Ballads of local states), *Bin feng* 豳風 (Ballads of the Bin 豳 region). The title of the poem is *qi yue* 七月 (The Seventh Month).

[12]Pang Pu 龐樸, "Houli gouchen" 火曆鈎陳 (The discovery of the Fire calendar), *Zhongguo Wenhua* 中國文化 (Chinese Culture), 1 (1989), pp. 1-35.

as we will see in the following sections.

The 28 lunar lodges (*xiu* 宿) belong to the most important structures in the Chinese sky. They were constellations (or asterisms) marked along the ecliptic belt of the celestial sphere to serve as reference points for the motion of sun, moon and planets. Thus they were essential in calendar making.

The origin of the 28 *xiu* is a problem which has long been studied and disputed and is without common conclusion. Both the Indian and Arabic worlds have a similar system of lunar lodges[13]. Was there a common origin? Who influenced whom? We will not try to answer these questions here[14].

The idea to use asterisms like the characteristic constellations of 28 lunar lodges, all situated close to and along the ecliptic, to mark the position of sun, moon or planets was so fundamental for any attempt to make a calendar that any civilization could have done this. The identical sequence of the seasons in all Asiatic countries led to similar names for corresponding lodges. International contacts smoothed out differences. The Chinese *xiu*-system might have developed as four groups of seven lodges centered about the four cardinal points.

The earliest complete set of the 28 *xiu* was discovered on a cover of a box excavated from the tomb of Zeng Hou Yi 曾侯乙 who died in 433 BC[15]. See Figure 2.1.

Almost all the *xiu* names appeared in the *Yue ling* 月令 (Monthly Ordinances) which was a part of the *Lüshi chunqiu* 吕氏春秋 (Master Lü's Spring and Autumn Annals) by Lü Buwei 吕不韋 who died in 235 BC[16].

---

[13]The Indian lunar lodges (*Nakṣatras*) were carefully described by E. Burgess (Translation of the *Sūrya Siddhānta*, Calcutta, 1935) and D. Pingree (History of Mathematical Astronomy in India, *Dictionary of Scientific Biography*, vol.15, New York 1978).

The Arabic lunar lodges (*Manazil*) date from about 800 AD and were initiated because of contacts with Indian astronomers . See P. Kunitzsch; "Al-Manazil", *Encyclopedia of Islam*, Vol.VI, Leiden 1991, pp. 374-376.

[14]For a general review on these problems see Chen Zungui 陳遵嬀, *Zhongguo tianwenxue shi* 中國天文學史 (History of Chinese Astronomy; vol.2, Shanghai, 1982), pp. 307-320.

[15]Wang Jianmin 王健民, et al, "Zenghouyi mu chutu de ershiba xiu qinglong baihu tuxiang" 曾侯乙墓出土的二十八宿青龍白虎圖象 (The twenty eight *xiu* and the images of Green Dragon and White Tiger discovered in the tomb of Zeng Hou Yi), *Wen wu* 文物 (Cultural Relics), 7 (1979), pp. 40-45.

[16]Lü Buwei was a minister of the Qin state during the later Zhanguo period. Formerly as a merchant he had protected the son (Ying Zheng 嬴政) of one of the princes. This son was

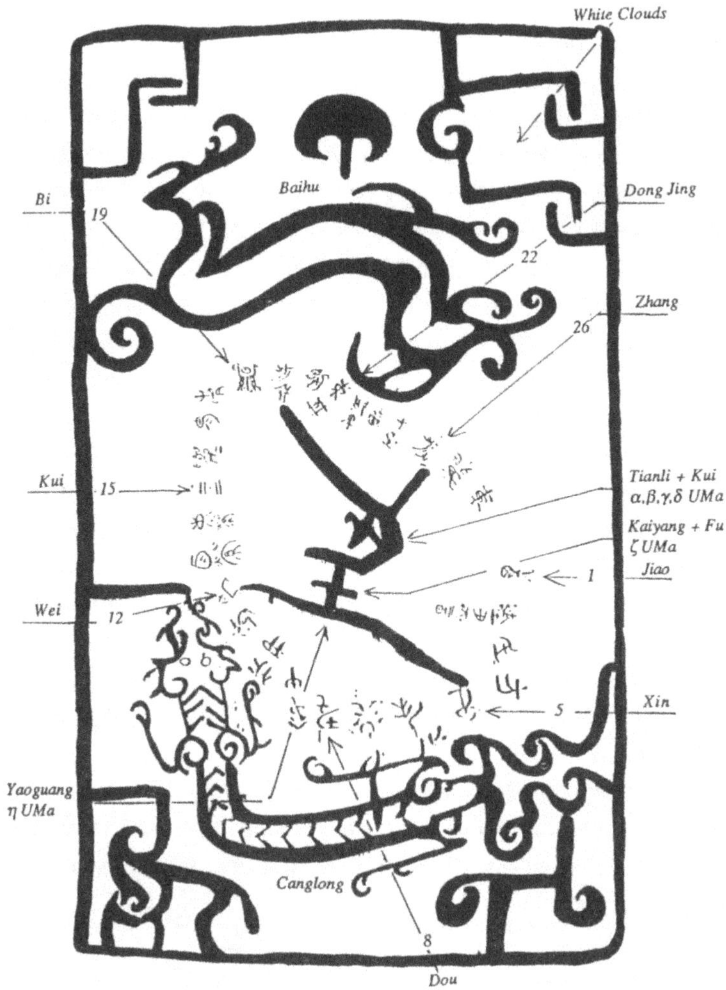

Figure 2.1: The names of 28 Lunar Lodges on the cover of a lacquered box from the tomb of Zeng Hou Yi 曾侯乙. Surrounding the big Chinese character *Dou* 斗 were the names of the 28 *xiu*. The character *Dou* represented the constellation *Bei Dou* 北斗. It was exaggerated in such a way that its strokes point to the four cardinal *xiu*, in this case to *Xin* 心, *Wei* 危, *Bi* 畢 and *Zhang* 張. On both ends of the cover were the images of Green Dragon (*cang long*) and White Tiger (*bai hu*). The date of the tomb is 433 BC. Some *xiu* names are indicated by us for understanding.

The *Yue ling* includes monthly celestial phenomena such as the sun's position among the stars, and the culminating stars at dusk and dawn. It was written in an even more systematic way than the *Xia xiao zheng*.

This is most of what we know about the Chinese constellations before the Qin and the Han times[17]. During the Zhan Guo period, there were two astronomers, Shi Shen 石申 (fl. 370 - 340 BC) and Gan De 甘德 (c. 3rd century BC), to whom any study of the Chinese sky might refer. The Chinese sky has been composed of constellations of three astronomical schools, two of which attributed them to these two astronomers. One may accordingly think that most constellations were named by Shi Shen and Gan De and therefore that the names already existed during the Zhanguo period[18]. But we would like to point out that it is naive to believe too many of the claims made in ancient literature without careful analysis. Quite a lot of Shi Shen's and Gan De's constellations are of later origin as we will discuss below.

2.2 *Stars in the* Tianguan shu

The earliest book which systematically described stars in the whole sky is the *Tianguan shu* 天官書, a Monograph on Celestial Officials (Constellations) written by Sima Qian 司馬遷 (145 - 87 BC). When he wrote his famous *Shi ji* 史記 (Historical Records), he devoted this special chapter to the description of stars. Sima Qian was an astronomer himself. He had the position of *taishi ling* 太史令 (Prefect Grand Astrologer) at the court of Han Wu Di 漢武帝. or Emperor Wu (140 - 87 BC) He organised appreciable astronomical activities preparing for the calendar reform during the reign of Emperor Wu. His father, Sima Tan 司馬談, also had the position of *taishi ling* before his son. We have every reason to believe that what Sima Qian said about the sky was from authoritative sources. Sima Qian and his father had learned about the sky from an astronomer whose name was Tang Du

to become the famous Qin Shi Huangdi 秦始皇帝, the First Emperor of the Qin dynasty.

[17]For a complete textual survey of the records of star names in ancient literature, see Wu Qichang 吳其昌 [1943].

[18]Wu Qichang, for example, held this opinion. See note above.

唐都 (c. 120 BC)[19]. In Tang literature we discovered material saying that Tang Du wrote several books on astronomy, which might still have existed during the Tang dynasty. One of them was a preliminary type of star map based on observations by means of an equatorial armillary sphere[20]. During Sima Qian's lifetime there were several famous astronomers. One of them was Luoxia Hong 落下閎 (fl. 140 - 104 BC) who invented the armillary sphere. The observation and description of the sky were already quite mature at Sima Qian's time. So we consider his descriptions in the *Tianguan shu* as the general view of the Chinese sky during the Former Han (Qian Han 前漢) or Western Han (Xi Han 西漢) (206 BC - AD 8) .

About 90 asterisms were mentioned in the *Tianguan shu*, including the 28 *xiu*. Their positions in the sky, relative to each other, were described and even the number of stars in the asterisms were given in most cases. This book registered 412 stars without including asterisms of which no number of stars was mentioned. Even the brightness of stars was given in words which might be considered as equivalent to magnitudes[21].

The *Tianguan shu* divided the sky into five *gong* 宮 (Palaces). The *zhong gong* 中宮 (Central Palace) indicated the area surrounding the celestial North Pole (*Beiji* 北極). The North Pole held the most dignified position in Chinese cosmology simply because of the fact that all other stars in the sky apparently moved around it. In the *Tianguan shu* the circumpolar area was established as the central imperial palace with star names like emperor, queen, princes, and all kinds of court officials. The other four palaces are the four cardinal palaces along the ecliptic, namely the East, South, West and North Palaces. Each of them represented the most prominent stars during one of the four seasons of the year. The 28 *xiu* were divided into four groups of seven *xiu* each group representing a palace.

In the *Tianguan shu* typical ideas reflecting the political system were presented like a central imperial court controlling the provinces. Celestial

---

[19]See the *Shi ji* 史記, ch.130, *taishigong zhixu* 太史公自序 (Autobiography of the Imperial Historian (Sima Qian himself)).

[20]Li Feng 李鳳 (c. 664) said in his *Tianwen yaolu* 天文要錄 (Records of the Most Important Astronomical Matters) that Tang Du of the Former Han had made a star map based on observations by means of an equatorial armillary sphere.

[21]Bo Shuren 薄樹人 "Sima Qian - woguo weidade tianwen xuejia" 司馬遷 – 我國偉大的天文學家 (The great Chinese astronomer Sima Qian), *Zhiran zhazhi* 自然雜志 (Nature), 4 (1980), No.9, pp.685-688.

control was realized by the movement of *Bei Dou* around the Central Palace. The text goes:

> *Bei Dou* serves as the chariot of the emperor and effectuates its control over the four cardinal points by revolving around the center; it separates the *yin* 陰 (feminine or negative force) and the *yang* 陽 (masculine or positive force) and regulates the four seasons; it maintains balance between the Five Elements (*wu xing* 五行); it regulates the moving of the celestial objects; it determines the epoch of the calendar.

We clearly see the general idea of constructing the sky as a celestial counterpart of the terrestrial imperial state, hence the book's title Celestial Officials (*tian guan* 天官). During the Han there apparently existed an academic tendency to correlate the sky with human society. The impetus came from the philosophy of *tian ren gan ying* 天人感應 (Interaction between Heaven and Man) which came to a climax during the Han with the writings of the famous philosopher Dong Zhongshu 董仲舒 (179? - 104? BC). Therefore more and more constellations were constructed in the sky to represent a real celestial society. The Signs in the Heavens could be interpreted straightforwardly in correlation with human affairs.

Between the time of Sima Qian and the middle of the Later Han or Eastern Han (AD 25 - 220) much work was done to describe and interpret the starry sky. Most of this work was done in the form of *wei shu* 緯書 ("woof" books)[22]. In the Tang *Kaiyuan zhanjing* 開元占經 (Treatise on Astrology of the Kaiyuan reign period) at least 70 kinds of *wei shu* from the

[22] *Wei shu* 緯書 should be understood in connection with *jing* 經 ("warp"). *Jing* usually meant the Confucian classics like the *Chun qiu* 春秋 (Spring and Autumn Annals) and the *Yi jing* 易經 (Book of Changes). *Wei shu* were the explanations of *jing*, reason why they were called "woof" and "warp" books respectively, inspired by the weaving technique. There was an academic tendency among Han-philosophers to expound their own philosophies in the form of explanations or commentaries on the Classics. Dong Zhongshu perfected his philosophy of *tian ren gan ying* in his *Chunqiu fanlu* 春秋繁露 which was written in the style of *wei shu*, although its content was highly influential and valuable in Chinese history of philosophy. Usually *wei shu* meant the literature that contained esoteric explanations. Nevertherless many astrological descriptions about the sky and sometimes advanced astronomical ideas could be found in *wei shu*. In *wei shu* we see an academic tendency to organise the sky into a counterpart of human society. *Wei shu* were popular during the reign of Han Cheng Di 漢成帝 (32 - 7 BC) and Ping Di 平帝 (AD 1 - 5).

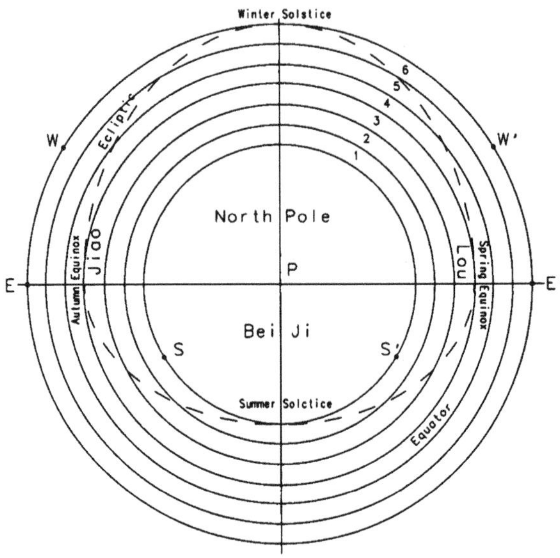

Figure 2.2: The "seven orbits and six belts" (*qi heng liu jian* 七衡六間) diagram of the *Zhoubi suanjing* 周髀算經. This diagram is designed by us according to the descriptions in the book.

Han were quoted. In "woof" books stars and constellations were frequently interpreted as presiding over terrestrial events in order to correlate them with celestial events.

The cosmology of the Han time developed to a stage which made a better description and even the mapping of the sky necessary. The cosmology of *gai tian* 蓋天 (Covering Sky) assumed its most refined form in the book *Zhoubi suanjing* 周髀算經 (Arithmetical Classic of the Gnomon and the Circular Paths). The sky was a circular cover which was divided into six circular belts by seven circles at different distances to the pole. The middle circle was the equator. This is the so-called *qi heng liu jian* 七衡六間 (seven orbits and six belts) diagram. See Figure 2.2. As the sun moved along the ecliptic it stepped into different belts so that the 24 *jie qi* 節氣 (solar periods) were produced[23]. This was mathematically a sophisticated

[23]For a thorough discussion of this diagram see Shigeru Nakayama, *A History of Japanese Astronomy: Chinese Background and Western Impact* (Cambridge Mas-

way to explain the seasons. It was said that the earliest star map was in the form of a planisphere which was called *gai tu* 蓋圖 (Map of the Covering Sky)[24].

Further development of cosmology during the Han led to the *hun tian* 渾天 (Enveloping Sky) theory with a celestial sphere. This theory coincided with the invention of the armillary sphere. A celestial globe was used to illustrate the daily motion of the sky. The distance of the stars to the pole could be measured by means of the armillary sphere and the sky could be precisely projected onto a celestial globe.

The time was ripe to study the starry sky. It is quite remarkable that almost at the same time in the Greek world, Hipparchus (c. 162 - 126 BC) and Ptolemy (AD 127 - 151) divided the sky into many constellations and measured the coordinates of most bright stars.

2.3 *Asterisms of the three astronomical schools*

As has been mentioned above the Chinese sky consists of constellations from three astronomical schools. These schools were called the three *jia* 家 (families) or the three *Shi* 氏 (families): *Shi Shi* 石氏, *Gan Shi* 甘氏 and *Wuxian Shi* 巫咸氏, as they were respectively named after three historical personages from before the Han: Shi Shen, Gan De and Wuxian.[25] Shi Shen 石申 and Gan De 甘德 were two astronomers from the Zhanguo period (481 - 222 BC), as we have said, while Wuxian 巫咸 was a legendary figure from the Yin dynasty. It is said that both Shi Shen and Gan De wrote books on astronomy and astrology, but apparently the original books were lost by Han times[26].

sachusetts, 1969), pp. 29-31.

[24]Qian Baocong 錢寶琮 made this suggestion through the study of *Zhoubi suanjing*. See "Gaitiansuo yuanliu kao" 蓋天說源流考 (On the origin of the *gai tian* theory), *Kexueshi jikan* 科學史集刊 (Studies on History of Science), 1 (1958), pp.29-46.

[25]It is understandable to call them Families in the view that astronomy was usually a kind of knowledge which traditionally passed from father to son. But during the Han the *Shi* should be understood as the three astronomical schools which continued the tradition of these three persons.

[26]It was mentioned in the *Shiji zhengyi* 史記正義 (Original Meaning of the *Shi ji*; by Zhang Shoujie 張守節 (c. 600 AD)) that Shi Shen had written eight chapters of *Tian wen* (Astronomy) and Gan De had also written eight chapters of *Tianwen xingzhan* 天文星占

As we looked at the constellations of the three schools, we found that quite a lot of them could not have existed before the Han. Some can easily be detected as direct deductions from descriptions in the *Tianguan shu*. This is true especially for the constellations of *Gan Shi* and *Wuxian Shi*. Thus we might say that the total system of constellations of the three schools was developed during the Han.

The constellations of the three astronomical schools were studied and conflated by Chen Zhuo 陳卓, an astronomer of the Sanguo 三國 (Three Kingdoms) period (AD 220 - 280). Of all constellations he identified 92 as belonging to *Shi Shi*, 118 to *Gan Shi* and 44 to *Wuxian Shi*. With the 28 *xiu* and one star (ζ Sco) in *Xin* 心 *xiu*, called *Shenggong* 神宮, it amounted to 283 constellations comprising of 1464 individual stars in the Chinese sky. It was said that Chen Zhuo had made a thorough study of the constellations of the three schools, made star maps out of them and added astrological commentaries in which he distinguished constellations of different schools[27].

The works of Chen Zhuo have long been lost. But fortunately all the information concerning the constellations of the three schools has been passed on to us through the astrological works of the Tang dynasty. We have the *Kaiyuan zhanjing* 開元占經 (Treatise on Astrology of the Kaiyuan reign period) by Qutan Xida 瞿曇悉達 (fl. 718) and transcripts of the astrological texts from the Dunhuang 敦煌 Manuscripts[28]. In both of them the constellations of the three schools were treated separately. Therefore no confusion is possible. These two primary sources for our study were not available to historians before the seventeenth century. The former one was discovered in the belly of a Buddha statue in the late seventeenth century.

(Astronomy and Star Prognostication). Portions of their work have been incorporated into the *Gan Shi xingjing* 甘石星經 (Star Manual of the Masters Gan and Shi). This book was basically made by Han astronomers. See Qian Baocong [1937].

[27] *Jinshu tianwen zhi* 晉書 • 天文志 (Treatise on Astronomy of the History of the Jin Dynasty). An English translation with comments was made by Ho Peng Yoke, *Astronomical Chapter of the Jin Shu* (Paris-The Hague, 1966).

[28] By Dunhuang Manuscripts we refer to the huge mass of medieval manuscripts which were discovered around the beginning of this century in Dunhuang, Western Gansu. Most of the manuscripts are preserved in the Stein collection of the British Museum in London and in the Pelliot collection of the Bibliotheque Nationale in Paris. The P 2512 manuscript in Paris contains astrological texts about the 28 *xiu* and the constellations of the three schools. The stars described in it are essentially the same as those in the *Kaiyuan zhanjing*.

Historians began to pay attention to it only as late as the 20th century. The Dunhuang Manuscripts were discovered in the beginning of this century.

In the *Kaiyuan zhanjing* the celestial coordinates of the leading stars, the *ju xing* 距星 of the 28 *xiu* and the constellations of *Shi Shi* were preserved. This is known to historians as the *Shi shi xingjing* 石氏星經 (Star Manual of Master Shi) Star Catalogue[29]. This catalogue is important not only because the date of the constellations of *Shi Shi* could thus be derived, but also because it provided the necessary data for the reconstruction of the constellations, as we will see in the next chapter.

The practice of distinguishing the constellations of the three astronomical schools was followed by astronomers after Chen Zhuo. When Qian Lezhi 錢樂之 (fl. 424 - 453), an astronomer of the South and North dynasties, made a celestial globe in the year of 436 AD, he used red, white and black colours to distinguish the constellations of the three schools: red for *Shi Shi*, black for *Gan Shi* and white for *Wuxian Shi*[30]. This was also practised when making star maps in later times.

2.4 *The Chinese sky during the Tang and Song*

Chen Zhuo's conflation of the constellations of the three astronomical schools provided a new version of the Chinese sky. It was inherited by the Tang astronomers and was established as the traditional Chinese sky. No remarkable changes happened thereafter, except for a few cases of introducing different star names (actually they were synonyms) or regrouping stars into new asterisms[31].

At the beginning of the Tang dynasty, a general description of the sky in the form of poetry was written by Wang Ximing 王希明 (about 6th century AD) which was called *Bu tian ge* 步天歌 (Song of Pacing the Heavens). In this popular work, the constellations mentioned were still the same as those of the three schools, but no longer distinguished according to which school

[29]see next chapter, page 39.

[30]See *Suishu tianwen zhi* 隋書 • 天文志 (Treatise on Astronomy of the History of the Sui Dynasty).

[31]Chen Meidong 陳美東, "Chen Zhuo xingguan de lishi shanbian" 陳卓星官的歷史嬗變 (The evolution of Chen Zhuo's constellations), *Kejishi wenji* 科技史文集 (Collected Studies on History of Science and Technology), vol.16, 1992, pp.77-91.

they had belonged to. The asterisms were described in the sequence of the 28 *xiu* (each of them representing an "hour-angle" segment of the sky) and three *Yuan* 垣, three areas enclosed by stars linked up like city walls, namely *Ziwei* 紫微 *Yuan*, *Taiwei* 太微 *Yuan*, and *Tianshi* 天市 *Yuan*. These wall-systems already existed among the constellations of *Shi Shi*, but they had not been emphasized as main structures of the sky. This "Three *Yuan* and Twenty Eight *xiu*" sequence of describing the sky was followed by most astronomers after the Tang.

Moreover there were two astronomical chapters of *Jin shu* and *Sui shu* both written by the famous Li Chunfeng (fl. late 7th century) of the early Tang dynasty. Also in these two books all the constellations of the three schools were mixed up with one another and fitted into a new way of describing the sky, namely, the sequence of the "Three *yuan* and Twenty Eight *xiu*". As far as the star names are concerned, there is not much to be said after the early Tang dynasty. More interesting were the star observations which were made in later times. Star catalogues and star maps were made from these observations and some of them have been preserved to the present time. They are valuable material for the study of the Chinese sky, especially for an effort to reconstruct the ancient Chinese constellations.

During the Kaiyuan 開元 reign period (713 - 741) of the Tang, the monk astronomer Yi Xing 一行 (683 - 727; his secular name was Zhang Sui 張隧) organized a large number of astronomical measurements made at widespread places. The coordinates for the determinative stars (*ju xing* 距星) of the 28 *xiu* were remeasured. New measurements were extended to other stars besides the 28 *xiu* and their new values were compared with the values in the "old classics" (*jiu jing* 舊經) as Yi Xing had them[32]. In chapters 110 to 116 of the *Kaiyuan zhanjing*, the difference between the newly measured positions and the positions of stars on the ancient star maps or in the "old classics" were discussed in detail. These discussions were based on the observations by Yi Xing. The differences were also recorded in the astronomical chapters of the *Jiu tang shu* 舊唐書 (Old History of the Tang Dynasty) and the *Xin tang shu* 新唐書 (New History of the Tang Dynasty).

Yi Xing's group also observed stars in the southern hemisphere which

[32]See *Jiu tang shu tianwen zhi* 舊唐書・天文志 (Treatise on Astronomy of the Old History of the Tang).

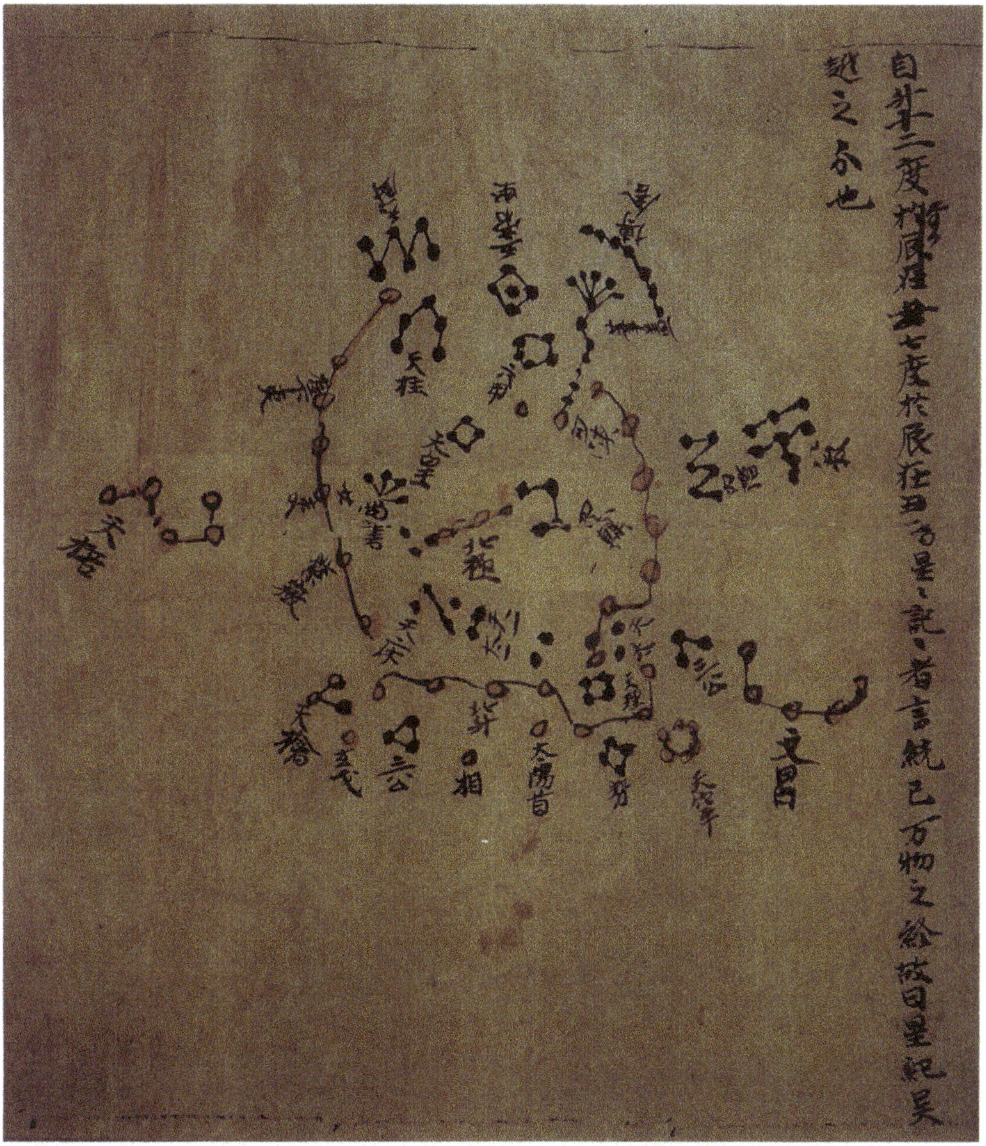

Star maps from the Tang dynasty, discovered in Dunhuang, Gansu. This part of the map shows the polar region of the sky. Constellations of the three Schools were distinguished with different colours: yellow and white for *Shi Shi* and *Wuxian Shi* (sometimes they are confused because of the colour of the paper); black for *Gan Shi*.

Permision of The Britisch Library (Ms S 3326).

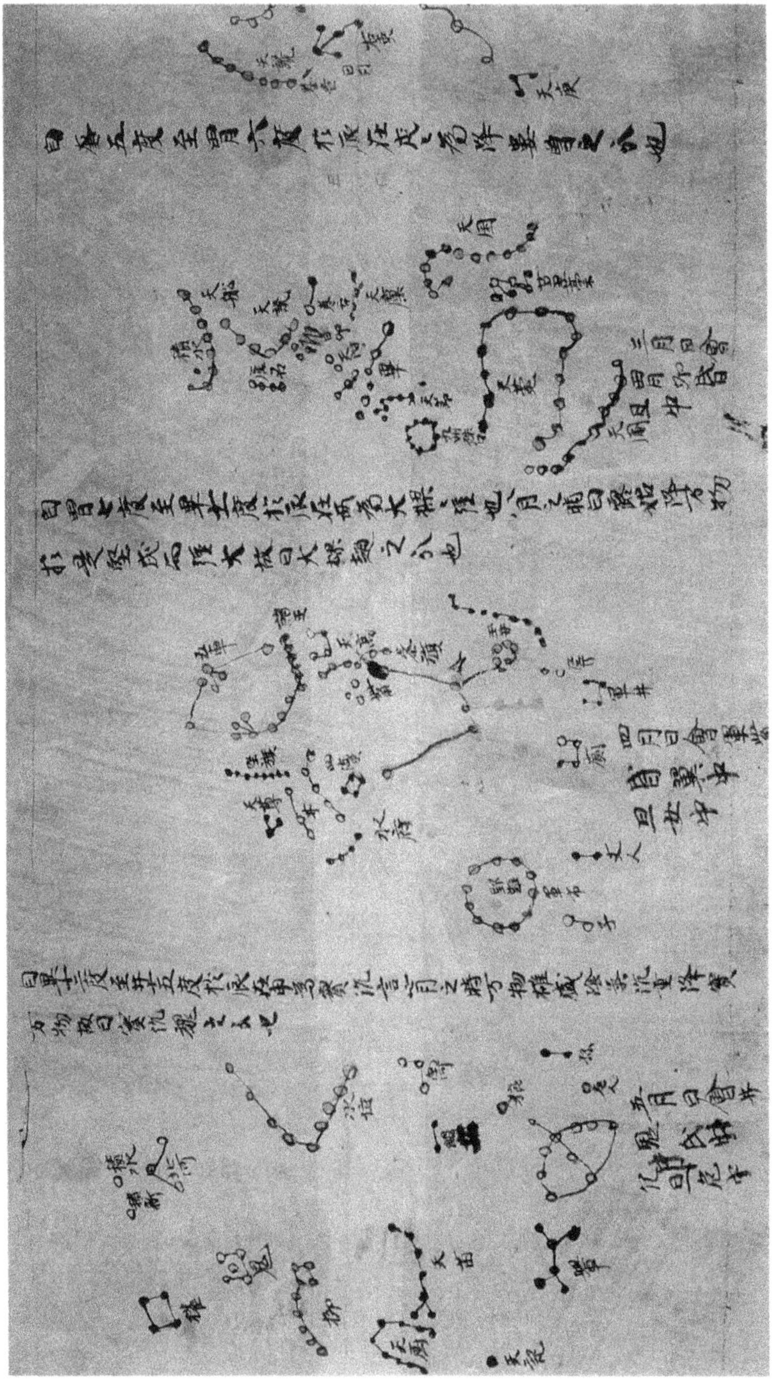

Star maps from the Tang dynasty, discovered in Dunhuang, Gansu. This part of the map shows the autumn section of the sky. Well-known constellations like *Shen* (Orion) and *Wuche* (Auriga) are shown.
Permission of The British Library (Ms S 3326).

had been considered by ancient astronomers as hidden and invisible below the horizon. "They went as far south as Jiao Zhou (modern Hanoi of Vietnam), from where the *Laoren* 老人 star was seen very high above the horizon, and below *Laoren* many more large and brilliant stars were visible. In former times they had never been named and charted. They were all observed to about 20 degrees from the southern celestial pole."[33] But their observations of these southern stars did not result in any remarkable modification of the old system of constellations; no new constellations were added. No star map exists today of Yi Xing's observations.

There exists a star map from the Tang. It was preserved in the Dunhuang Manuscripts discovered around the beginning of this century. This is certainly the oldest extant star chart of China[34]. This star chart includes the north polar area and twelve hour-angled segments of the sky. The Constellations are drawn in three colours, white, black and yellow, to correspond with the Three Astronomical Schools. Although the star chart itself is by no means an accurate projection, the star names, number, and their relative positions are still valuable for the purpose of tracing the constellations of the three schools.

Then came the Song 宋 dynasty, during which science and technology were developed to an extremely high level. There were tremendous efforts in inventing and making astronomical intruments, always accompanied by new astronomical measurements. From historical sources we know that during the Song there were seven periods of observation of the stars; three of them concerned all the constellations in the sky, not only the 28 *xiu*. They were:

— the measurements made during the Jingyou 景祐 reign period (1034 - 1038) by Yang Weide 楊維德;

— the measurements made during the Huangyou 皇祐 reign period (1049 - 1054) by Zhou Cong 周琮;

— the measurements made during the Yuanfeng 元豐 reign period (1078-1085).

Coordinates of stars from the Jingyou and the Huangyou observations can still be found in the Song dynasty documents. The Jingyou Star Catalogue

[33] In the *Jiu tang shu tianwen zhi*.

[34] Needham dated it 940 AD. See *Science and Civilization in China*, section 20, p.264.

and the Huangyou Star Catalogue, as we call them, could be reestablished from the ancient books. This job was excellently done by Pan Nai[35]. The Jingyou Star Catalogue consists of coordinates of 341 stars and the Huangyou Star Catalogue of 360 stars. Both catalogues give the coordinates of the leading stars of all 283 constellations and some individual stars in a number of prominent constellations such as *Beidou* 北斗, *Tianshi Yuan* 天市垣 etc., hence 341 and 360 stars. These two catalogues for the first time gave the data for the constellations of *Gan Shi* and *Wuxian Shi* in addition to those of *Shi Shi.* They will be used for the purpose of reconstructing Gan De's and Wuxian's constellations in chapter 4. According to Pan Nai's analysis, the Huangyou catalogue was more precise than the Jingyou catalogue[36]. So we will use the Huangyou data in cases of doubt.

There also exist two star charts which are valuable for the study of the Chinese sky of the Song dynasty. One is the set of star maps included in the *Xin yixiang fayao* 新儀象法要 (New Design for Astronomical Instruments) by Su Song 蘇頌 (1020 - 1101). Su Song was the famous astronomer who invented and built the *shui yun yi xiang tai* 水運儀象臺, a tower-clock observatory powered by flowing water. The book was a general description of the principle of the observatory. The observatory was constructed in the year 1090 AD, and the book was written right after this. The star maps are five in number, one of the north polar region or *Ziwei Yuan*, two cylindrical orthomorphic cartesian projections of the ecliptic- and equatorial regions and two polar projections of the northern and the southern hemispheres. The south polar region was left blank because no observations were available at that time. The constellations of the three schools were also distinguished, but this time small circles were used to present stars of *Shi Shi* and *Wuxian Shi*, whereas black dots were used to present stars of *Gan Shi.* The positioning of the stars was carefully done and therefore the maps were quite accurate projections of the sky. No doubt the accurate measurements of stars made during the Song were used in making these maps[37]. On the maps all of the 1464 stars of 283 constellations were indicated. Thus the maps are quite

[35]Pan Nai 潘鼐, *Zhongguo Hengxing Guance Shi*, ch.6, pp.210-227.

[36]Pan Nai, *ibid.* ch.6, pp. 165ff. He even made a reconstruction of the Song sky based on the Huangyou catalogue.

[37]Pan Nai has suggested that the observations made during the Yuanfeng period (1078-1085) have been used for making the maps. *Zhongguo hengxing guance shi*, pp.241-253.

valuable for the reconstruction of the ancient constellations.

The other important star chart from this period is the well-known *Suzhou tianwen tu* 蘇州天文圖 (Suzhou Planisphere) which was engraved in stone and which still exists as a stele in the Confucian Temple (*wen miao* 文廟) in Suzhou. The planisphere was prepared by Huang Shang 黄裳 (1147 - 1195) in 1193[38] and was committed to stone by Wang Zhiyuan 王致遠 in 1247[39]. It presents the sky up to 50 degrees south of the celestial equator, the part which was visible in central China. The ecliptic and the equator are presented with intersecting circles whereas radiating lines connect the north pole and the determinative stars of the 28 *xiu* as shown on the planisphere. In this way stars could be engraved according to their coordinates. All the constellations of the three schools are presented on this one single map. The number of stars which still can be recognized amounts to 1436, not too different from the traditional number of 1464[40]. Investigations have shown that the position of the stars on the map is rather accurate[41], which is understandable in view of the high quality observations of stars during the Song dynasty. The planisphere can be recognized as a scientific star map and therefore serves as an important landmark from where we can recover the Chinese sky before the Song.

2.5 *Star observations during the Yuan, Ming and Qing*

Like traditional Chinese science and technology in general, the observation of stars reached its peak during the Song. The star catalogues and star maps from the Song provided us with the most valuable material concerning the traditional Chinese constellations. Thereafter the material was not so abundant. But during the following Yuan, Ming and Qing dynasties astronomical observations were continually carried out. Influences from the

[38]The *Song shi Huang Shang zhuan* 宋史・黄裳傳 (Biography of Huang Shang of the History of the Song).

[39]This is known from the inscriptions below the chart.

[40]The inscriptions accompaning the map give the number 1564, which is obviously a mistake. It should be 1464.

[41]Du Shengyun 杜升雲, "Suzhou shike tianwentu hengxing weizhi de yanjiu" 蘇州石刻天文圖恆星位置的研究 (On the position of the stars on the Suzhou astronomical chart), *Beijing shifan daxue xuebao* 北京師範大學學報 (Journal of Beijing Normal University), 1982, No.2, p.81.

outside world, especially from the Islamic world during the Yuan and the Ming, and from the West during the later Ming and Qing, were getting constantly stronger. These have had their impact not only on the observation of the stars as a scientific practice, but also on the system of the Chinese sky itself. New methods of observation were introduced and star catalogues became more accurate and were enlarged. This certainly helps to identify stars in the ancient constellations, since all earlier star catalogues only presented coordinates of a limited number of stars, as we have seen above. In the meantime, these influences also caused some changes in the Chinese constellations themselves. It happened that some star names were connected with stars different from those of older times because of new identifications made according to some foreign star catalogues. This happened especially after the introduction of the western knowledge of stars as we will see below. Therefore we do not consider the sky which passed down to us through these dynasties as the real traditional Chinese sky. We will only make use of it in our effort to reconstruct the ancient sky.

As a matter of fact there was some influence from the Islamic world during the Yuan. During the early Yuan dynasty Guo Shoujing 郭守敬 (1231 - 1316) built thirteen astronomical instruments to get data for the purpose of making the *Shou shi* 授時 Calendar. Some large instruments, the *jian yi* 簡儀 (Simplified Armillary) for example, were probably made under the influence of Persian or Arabic sources[42]. More interesting for us is Guo Shoujing's observation of stars during his calendar making. It is said in the *Yuan shi* 元史 (History of the Yuan Dynasty) that Guo Shoujing measured the 28 *xiu* and all constellations and compiled a star catalogue from his observations[43]. Moreover, he also measured unnamed stars for the first time, and compiled a star catalogue of unnamed stars. This interest in stars beyond the Chinese tradition could only have been triggered by certain foreign star catalogues which contained different stars. Unfortunately Guo Shoujing's data have not been preserved. Only the data about the 28 *xiu* were recorded in the *Yuan shi* and continued to be used during the succeeding

[42]For a discussion on the Arabic influence on instrumentation at Guo Shoujing's time see Needham, *Science and Civilization in China*, section 20, p.372.

[43]The title of the star catalogue was the *Xince ershiba xiu cazuo zhuxing ruxiu quji* 新測二十八宿雜坐諸星入宿去極(Newly Measured Equatorial Coordinates of the Stars of the 28 *xiu* and the other Constellations).

Ming dynasty[44].

It is generally agreed that the Ming period did not contribute much to traditional astronomy. It is believed that Guo Shoujing's calendrical system was simply adopted and that there were no serious improvements or inventions. But the discovery of a star catalogue from the Ming time may change this common view. The star catalogue, entitled *Sanyuan lieshe ruxiu quji ji* 三垣列舍入宿去極集 (Collection of Coordinates of Stars in the Three *Yuan* and the Twenty Eight *Xiu*), was discovered in a Ming transcript with the general title *Tianwen huichao* 天文彙鈔 (Collection of Astronomical Material); thus we call it the *Tianwen huichao* Star Catalogue[45]. The star catalogue was presented in quite an unusual way; its coordinates were noted on many pages of illustrative star maps with data attached to the symbols (small circles) of stars. The catalogue is not intact but still the coordinates of 740 stars have been preserved. We know that it gave coordinates for every individual star of the constellations, not only the leading stars or the determinative stars. Some researchers have suggested that this catalogue might contain the observations made by Guo Shoujing because the *Yuan Shi* mentioned his catalogue with a similar title[46]. However, our recent study has shown that the observation time was at the beginning of the Ming dynasty[47]. We know from the *Ming shi* 明史 (History of the Ming Dynasty) that there were a great deal of astronomical activities in the new capital Nanjing at the beginning of the Ming; so it was quite probable that they measured the star positions. At that time Islamic astronomy had a strong influence in China. A special department, the *Hui-hui ke* 回回科 (Islamic Section), was set up to translate and study the Persian and Arabic astronomical works. The Ptolemaic star catalogue was no doubt known to the Ming astronomers[48].

[44]The Ming *Da tong* 大統 (Grand Unification) calendar system kept using Guo Shoujing's values for the widths in right ascension of the 28 *xiu*. See the *Ming shi tianwen zhi* 明史・天文志 (Treatise on Astronomy of the History of the Ming Dynasty).

[45]Pan Nai, *Zhongguo Hengxing Guance Shi*, pp.274-277.

[46]Chen Ying 陳鷹 [1986] and Pan Nai [1989] have made independent studies of the catalogue and came to the same conclusion. At the beginning of their analysis both of them assumed that the star catalogue was compiled by Guo Shoujing.

[47]Sun Xiaochun, Bo Shuren, J.Kistemaker, "The *Tianwen huichao* star catalogue and the early Ming astronomy," *Oriental Astronomy from Guo Shoujing to King Sejeong* (Proceedings of the Seoul Conference, October,1993), in press. We determined the date of the star catalogue by analysing the *qu ji du* of the stars using Fourier analysis.

[48]Among the translated Arabic works, one was called *Mingyi tianwen shu* 明譯天文書

And that could have easily stimulated observations for larger catalogues. The *Tianwen huichao* Star Catalogue gives data as precise as one tenth of a degree and the accuracy is up to half a degree. Thus it provided us with another valuable source of data to study the constellations.

The Western knowledge of astronomy was introduced into China during the later Ming by the Jesuit missionaries. In the early seventeenth century there was a great enthusiasm in translating western astronomical works. In 1629 Xu Guangqi 徐光啓 (1562 - 1633) was appointed to organize the calendar reform in which the whole system of Western astronomy was to be introduced. A little earlier, the Western sky had already been introduced into China in the *Jing tian gai* 經天該 (Rhymed Description of Stars) by Li Zhizao 李之藻 (1562 - 1633) with the collaboration of Li Madou 利瑪竇 (Matteo Ricci, 1552 - 1610). The stars of the Western constellations were described in the style of poems just like in the Tang *Bu tian ge* 步天歌 (Song of Pacing the Heavens). Some new constellations were made out of Western constellations in the southern hemisphere and were added to the Chinese system. When the calendar reform started, Xu Guangqi made a large effort to measure precisely the positions of the fixed stars. In doing so he compared some Western star atlasses and catalogues. The coordinates of traditional Chinese stars were adjusted to those in the Western catalogues. The precise western coordinates could be readily tranformed to the required epoch in order to make a new star catalogue for the Chinese purpose. This work was done excellently by the Jesuit missionary Tang Ruowang 湯若望 (Adam Schall von Bell, 1591 - 1666) and his collaborators at the Astronomical Bureau during the calendar reform. A star catalogue was compiled and finally included in the *Chongzhen lishu* 崇禎曆書 (Astronomical Encylopaedia of the Chongzhen Period) which represented the entire achievement of the calendar reform[49]. In the star catalogue the ecliptical and equatorial coordinates of 1365 stars were given for the epoch of 1628, the first year of the Chongzhen 崇禎 reign [50]. Among them were 126

(Astronomical Book Translated [by Imperial Order of] the Ming) in which a list of 30 stars for calendrical purposes was given with sequential numbering according to the Ptolemaic catalogue in the *Almagest*.

[49]This star catalogue is known as Nr. 137 of the publication of *li jü* 曆局 the Bureau of Calendar.

[50]The *Chongzhen lishu* star catalogue used Right Ascension and Declination instead of Chinese *ru xiu du* 入宿度 (Degrees into the *xiu*) and *qu ji du* 去極度 (Distance to the

stars in the southern circumpolar region which had never been registered by the Chinese before. So 23 new constellations were added to the traditional Chinese system, the names of which were more or less a translation of the corresponding Western names of these constellations.

The traditional Chinese constellations were treated quite scientifically. If the stars could not be identified, then no coordinates were given for them, only the star names were listed. Some rather bright stars which had not been treated in the old Chinese system were included in the catalogue as *zeng xing* 增星 (supplementary stars). Thus this star catalogue can be regarded as the first correlation between the Chinese and the Western sky.

A group of star maps was made during the calendar reform. The most important one was the *Chidao nanbei liangzong xingtu* 赤道南北兩總星圖 (General Star Atlas of the Southern and Northern Hemispheres Divided by the Equator), copies of which are still preserved in the Vatican and in Beijing. The atlas was based on the *Chongzhen lishu* star catalogue, but additional measurements were probably made of stars which had no data in the catalogue because the number of stars drawn in the atlas totaled 1812 as Adam Schall von Bell (Tang Ruowang) indicated in the postscript of the atlas. The star atlas itself was a new version of the Chinese sky of the later Ming time. In a sense it re-shaped Chinese constellations to correspond better with Western constellations.

During the Qing dynasty there were continual efforts by missionaries working in the Qing Astronomical Bureau (*qin tian jian* 欽天監) to improve the star catalogue and the star atlas of the Chong Zhen period. More recent, and therefore more precise and larger, star catalogues from the West were used in attempts to identify all the traditional Chinese stars[51]. The Belgian missionary Nan Huairen 南懷仁 (Ferdinand Verbiest, 1623 - 1688) expanded the star catalogue up to 1876 stars covering the 1812 stars in Tang Ruowang's star atlas. This catalogue was collected in the *Lingtai yixiang zhi* 靈臺儀象志 (On the Astronomical Instruments in the Imperial Observatory) which he prepared during his service in the Astronomical Bureau. In

Pole); the 28 *xiu* system was abandoned. The ecliptical coordinates were given in the same way as in the Western star catalogue; only the twelve *ci* 次 (Jupiter's Mansions) were used instead of the twelve Zodiacs.

[51] In fact some Chinese constellations were impossible to identify because they had been constructed from imagination of *Gan Shi* and *Wuxian Shi*.

1744, the German missionary Dai Jinxian 戴進賢 (Ignatius Kögler,1680-1746) came to improve Nan Huairen's work. The *Yixiang kaocheng* 儀象考成 (General Description of Imperial Instruments) star catalogue was published by the Astronomical Bureau with Ignatius Kögler (Dai Jinxian) as chief editor. Flamsteed's star catalogue, much larger and more accurate, was consulted. The *Yixiang kaocheng* contained 3083 star positions, of which 1876 were numbered (300 constellations). This catalogue is the principal source nowadays for Chinese constellations.

Recently Yi Shitong [1981] made a correlative star map of the Chinese sky with the modern universal sky based on the *Yixiang kaoheng* star catalogue. This map is widely accepted and certainly valuable for studying the Chinese sky. But we should always keep in mind that this is not exactly the ancient Chinese Sky, it had already been modified during the Ming and the Qing by the work of the missionaries. Fortunately, we have all the historical information and material to trace back in history to see what the ancient Chinese sky was like in the Han time. This is one of the purposes of our present work.

CHAPTER THREE

CONSTELLATIONS OF SHI SHI

3.1 Shi Shi *and the* Shi shi xingjing

First let us emphasize that by *Shi Shi* we refer to the astronomical school which followed the tradition started by Shi Shen[1]. Shi Shen, according to Sima Qian, was an astronomer of the State Wei 魏 of the Zhangguo period[2]. It is said that he had written eight chapters of *Tian wen* 天文 (Astronomy)[3]. This book certainly existed in the early Han and as usual it was called *jing* 經, the Classic. It was probably also available to Sima Qian because he quoted it when discussing the movements of the planets in his *Tianguan shu*. The *jing* of Shi Shen (and also of Gan De whom we will discuss in the next chapter) was even more frequently quoted in astronomical works after the Former Han, such as the *Han shu tianwen zhi* 漢書・天文志 (Treatise on Astronomy in the *Han shu*[4]) and the *Xu han shu lüli zhi* 續漢書・律曆志 (Treatise on Harmonics and Calendrical Science in the *Xu han shu*[5]). But all these so-called "quotations" were not really from the original *jing*. New astronomical ideas and knowledge of the Han time were recognizable in these "quotations"; that means the *jing* was already developed and supplemented by Han astronomers. This was the common practice of ancient Chinese scholars who expressed their own ideas by enlarging the Classics. The result was that no original *jing* existed any more. The "quotations"

---

[1]In the text *Shi Shi* will also be referred to as Shi Shen's school. The same applies to *Gan Shi* and *Wuxian Shi*.

[2]See the postscript of the *Tianguan shu* in the *Shi ji* by Sima Qian.

[3]See footnote on page 25.

[4]The History of Han Dynasty.

[5]The Continued History of the Han Dynasty.

from the *jing* could be descriptions of current knowledge.

When "quoting" Shi Shen the text would start with "*Shi Shi yue* 石氏曰 (Master Shi said)" or "*Shi Shi jing yue* 石氏經曰 (the *jing* of Master Shi said)". The book title *Shi shi xingjing* 石氏星經 (Star Manual of Master Shi) was mentioned for the first time in the *Xu han shu lüli zhi*. The text goes like this: "*Shi shi xingjing* said: ` if measured along the ecliptic, *qian niu chu* 牽牛初 (the winter solstice[6]) is twenty degrees into *xiu Dou* 斗.'" Simple calculation of precession immediately shows this material was from the later half of the Former Han[7]. What *Shishi xingjing* says was not from Shi Shen's time at all. This example illustrates how the *Shishi xingjing* must be understood. It contained a body of knowledge which has been continuously updated and which was claimed to be from Shi Shen.

Shi Shi has played an important role in astronomy of the Han time. Astronomical discoveries were attributed either to *Shi Shi* or to *Gan Shi* at that time. The group of astronomers who were working during the Taichu 太初 calendar reform[8] were probably members of the *Shi Shi*. The introduction of the ecliptic (*huang dao* 黄道) into astronomical measurements was made by *Shi Shi*. In the second year of the reign of Ganlu (52 BC) Geng Shouchang 耿壽昌 suggested to use an instrument called *yuan yi* 圓儀 in a Memorial to the throne. With this instrument the solar and lunar positions were to be measured along the ecliptic and not along the equator as in former practice. The *Shi shi xingjing* discussed the same subject. But it was as late as the fifteenth year of the Yongyuan 永元 reign (103 AD) that the widths of the *Xiu* were really measured along the ecliptic. This was one of the achievements by *Shi Shi*. Remarkably enough, in the second year of the reign of Yuanhe 元和 (85 AD) an imperial edict was specifically issued saying that the system of *Shi Shi* was indispensable for astronomical studies (*Shi bu ke li* 石不可離). This shows how predominant *Shi Shi* was at that time.

During the Later Han there were some astrologers who began to at-

---

[6]*Qian niu chu* means the beginning of *xiu Niu* 牛 (or *Qianniu* 牽牛). Because the ancient *si fen* 四分 calendar determined the winter solstice at the beginning of *xiu Niu*, *qian niu chu* became a synonym of the winter solstice (*dong zhi* 冬至.)

[7]Qian Baocong 錢寶琮 dated it 70 BC, see page 43 below.

[8]A recent paper on the Taichu calendar reform is written by C. Cullen; "Motivations for scientific change in ancient China: Emperor Wu and the grand inception of astronomical reforms of 104 BC," *Journal for the History of Astronomy*, xxiv (1993), pp.185-203.

tribute their work to Wuxian; thus three main astronomical schools were active then. After the collapse of the Han empire, China was split up into warring states. That provided a fertile soil for all kinds of astrological publications. Again most of these works were attributed to the three schools. The result was that the *jing* of Shi Shen was even further enlarged. The original *jing* was no longer recognizable. But still these later books continued to "quote" Shi Shen.

There also exists another book entitled *Gan Shi xingjing* 甘石星經 (Star Manual of the Masters Gan and Shi.) We know this book was compiled by Ma Xian 馬顯 (c. 579 AD) as an appendix to a calendar treatise made by him[9]. As explained above, it "quoted" so much of *Shi Shi* and *Gan Shi* that an innocent reader could be easily mislead to regard it as the work of Shi Shen and Gan De together. Indeed, some editions of this book were claimed to be written by "Gan Gong 甘公 (Gan De) and Shi Shen from the Han". This attribution is ridiculous because Gan De and Shi Shen lived in the Zhanguo period about 300 BC. It can only imply that most of the content of the book was from the Han time.

The *Kaiyuan zhanjing* of the Tang "quoted" prodigiously from *Shi Shi.* For convenience historians simply gave all these "quotations" a collective title "*Shi shi xingjing,*" the Astrological Treatise of Master Shi Shen (hereafter to be abbreviated as *SSXJ*). It should not be mistaken for the original *Shi shi xingjing* as mentioned in the *Xu han shu lüli zhi*. The *SSXJ* is not a book, it is a title used by historians to refer to all astronomical source material from Shi Shen's time up to the Tang. It was attributed to Shi Shen. The most interesting material for our present study is that from the Han period.

3.2 The SSXJ *star catalogue*

Chen Zhuo identified 92 constellations as belonging to *Shi Shi*, 118 constellations as belonging to *Gan Shi* and 44 constellations to *Wuxian Shi*. This allocation has been accepted by astronomers of later times. In the *Kaiyuan zhanjing* the constellations of the three schools and the 28 *xiu* were dealt with most thoroughly in the astrological sense. Moreover, for the constella-

[9]The proper title of the book should be *Tongzhan daxiang li xingjing* 通占大象曆星經, the General Star Manual affiliated to the calendar of *Daxiang*. (Northern Zhou of 557 to 581 AD.)

tions of *Shi Shi* and the 28 *xiu*, the *Kaiyuan zhanjing* provided coordinates for their leading stars. This has become known as the *SSXJ* star catalogue. The study of this star catalogue is critical for the study of the constellations of *Shi Shi*.

3.2.1 The star catalogue preserved in the *Kaiyuan zhanjing*

Information about the constellations of *Shi Shi* can also be found in many ancient astronomical texts since the Han. For example, in the above mentioned *Ganshi xingjing* were detailed descriptions of the relative positions as well as of the astrological meaning of the constellations of *Shi Shi*. But the book gave coordinates of only a few stars. The complete quotation of the star catalogue of *SSXJ* is in the *Kaiyuan zhanjing* (hereafter to be refered as *KYZJ*)[10]. The data can be found in notes written in small characters between lines of the text on stars. See Figure 3.1. There exist other Tang-time references containing some data of the catalogue. The most important one is the *Tiandi ruixiang zhi* 天地瑞祥志 (Records of Auspicious Phenomena in Heaven and on Earth) by Sa Shouzhen 薩守眞 in the year 666 AD. The data available in this book can be used to check or even make up deficiencies in the data in the *KYZJ*.

The *KYZJ* text concerning the stars of *Shi Shi* begins with "Shi Shi said" or "*SSXJ* said." It contains the following information:

1. Names of the constellations; number of individual stars in each constellation; position of each constellation relative to other constellations or to some celestial features such as the Milky Way; the astrological protasis of these star names.
2. Description of the relative position of the leading star in each constellation. In most cases a star on the western boundary of the constellation was chosen. Exceptions could be stars of special astronomical or astrological meaning.
3. The *ru xiu du* 入宿度 (distance to the western boundary of a *xiu*) and the *qu ji du* 去極度 (distance to the pole) for the leading stars of each

[10]Complete as we call it, there are still six sets of coordinates missing. But fortunately four sets of the missing ones can be retrieved in another Tang book, the *Tiandi ruixiang zhi*. See below.

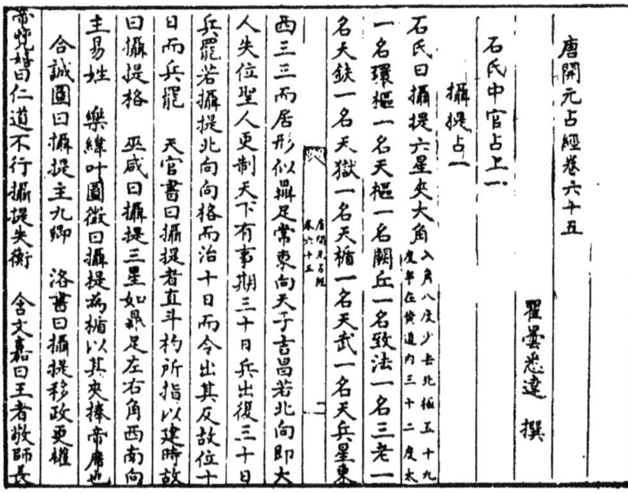
唐開元占經卷六十五
瞿曇悉達　撰
石氏中官占上一
攝提占一
石氏曰攝提六星夾大角 入角八度少去北極五十九度半在黄道内三十二度太
一名環樞一名天樞一名闕丘一名致法一名三老一
名天銊一名天獄一名天楯一名天武一名天兵星東
唐開元占經卷六十五　二
西三三而居形似鼎足常東向天子吉昌若北向即大
人失位聖人更制天下有事期三十日兵出復三十日
兵罷若攝提北向向格而治十日而令出其反故位十
日而兵罷　天官書曰攝提者直斗杓所指以建時故
曰攝提格　巫咸曰攝提三星如鼎足左右角西南向
主易姓　樂緯叶圖徵曰攝提爲楯以其夾擁帝席也
合誠圖曰攝提主九卿　洛書曰攝提移政更雄
帝覽嬉曰仁道不行攝提失衡　含文嘉曰王者敬師長

Figure 3.1: A page from the *Kaiyuan zhanjing* 開元占經. The smaller characters are "quotations" from the *SSXJ* star catalogue about the coordinates of the leading star of a constellation.

constellation, hereafter to be abbreviated as RXD and QJD, respectively. The RXD is astronomically equivalent to the Right Ascension of a star in an equatorial coordinate system. The only difference is that it does not count from the spring equinox, but from the determinative stars of the *xiu*. The RXD is always in the form of how many degrees in which *xiu*. Ancient Chinese astronomers used the north pole and the 28 determinative stars of the *xiu* to form a spoke-like set of meridian circles on the celestial sphere which served as an equatorial coordinate system. See Figure 3.2. The QJD is the complement of the Declination. The text did not give the RXD of the determinative stars of the 28 *xiu* since in this coordinate system they were always equal to zero.

4. The *huang dao nei wai du* 黄道内外度 (HDD). These coordinates are usually called Latitudes. They give the shortest distance from a star to the celestial ecliptic. It is not clear from the text how the Han astronomers could measure this distance with their primitive instruments. Perhaps a mechanical model like a globe was used. Yabuuchi Kiyoshi 藪内清 [1937] suggested that they measured pseudo-

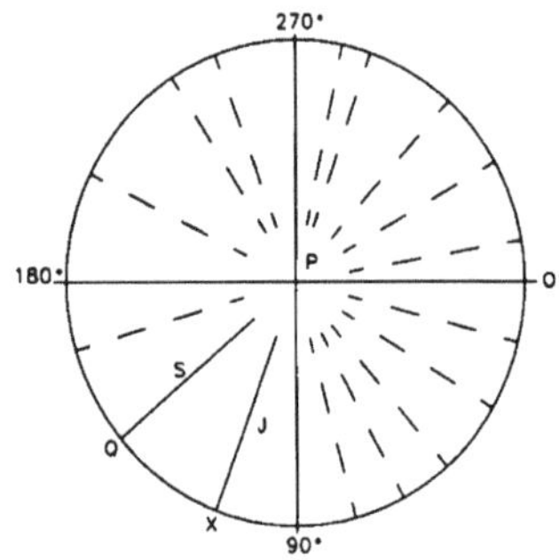

Figure 3.2: An illustration of the ancient Chinese equatorial coordinate system with the 28 *xiu*. They divide the celestial sphere into 28 sectors like an orange. QJD is the distance SP of the star S to the pole; its RXD is the angle QPX.

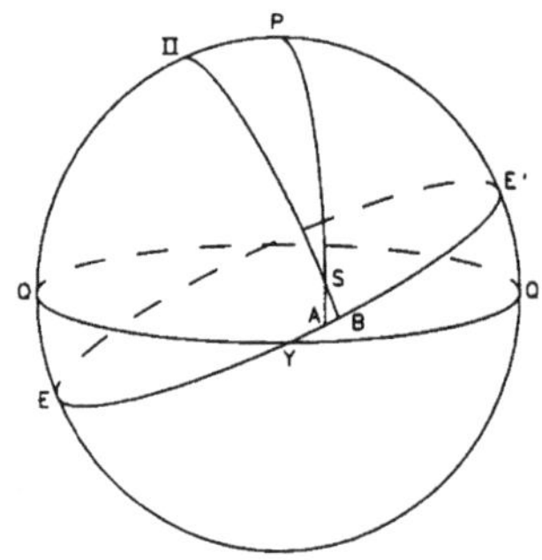

Figure 3.3: An explanation of the Pseudo-Latitude. With Π and P representing the ecliptical and equatorial poles respectively, the real Latitude of a star S is SB, while the Pseudo-Latitude is SA.

Latitudes along the meridian in the equatorial system[11]. See Figure 3.3.

3.2.2 Former studies on the *SSXJ* catalogue

The coordinates of the leading stars of these 120 constellations in the sky are extremely important for the identification of these constellations; 28 of these leading stars correspond to the determinative stars (*ju xing* 距星) of the Chinese lunar lodges (*xiu* 宿). These *ju xing* were used for time determination and have always been measured with special care. Former studies have considered the observation time of the data in the *SSXJ* star catalogue. They led to the following three conclusions:

- The time of observation would have been during the Zhanguo period.

[11] Yabuuchi Kiyoshi 藪内清, "To kaigen senkyō chū no seikyō" 唐開元占經中之星經 *Tōhō Gakuhō* (Kyoto), 8 (1937), pp.56-74.

Shinjō Shinzō [1933] used information on the position of Jupiter as quoted in the astronomical chapter of the *Han shu*. He concluded that the astronomical observations of Shi Shen were done in the middle of the Zhanguo period, about 360 BC. Furthermore, he argued that while observing the five planets, Shi Shen also measured the QJD of the 120 stars[12].

- There might have been two epochs.

By means of graphic analysis of the spread of the data, Ueda Yuzuru 上田穰 [1929] concluded that about one half of the data was observed in 360 BC, the other half in 200 AD[13].

Recently Pan Nai [1989] first assumed the data of the catalogue to be exact and then calculated a corresponding date for every star. Then he divided the catalogue into two parts, one with the QJD corresponding with dates from before Christ, the other with dates from after Christ, and made average dates for each set. They were 450 BC and 160 AD. He suggested that a complete star catalogue of Shi Shen existed aoubt 450 BC, but part of it was lost. Then astronomers of the Later Han might have made new observations to compensate for the lost part[14].

- The observations were made during the Former Han.

Yabuuchi Kiyoshi[15] and Qian Baocong[16] did independent textual studies and both concluded that the most important part of *SSXJ* was composed during the Former Han. Yabuuchi also used information concerning the position of the winter solstice and came to the conclusion that *SSXJ* contained the astronomical knowledge of 70 BC. According to him the observation of the star catalogue dates from that same epoch.

---

[12]Shinjō Shinzō, *Dongyang Tianwenxue shi yanjiu* 東洋天文學史研究 (Researches in the History of Astronomy in the Far East; Tokyo, 1928; Chinese translation by Chen Zungui, Shanghai, 1933.), pp.18-20.

[13]Ueda Yuzuru 上田穰, *Seki-shi Seikyō no Kenkyū* 石氏星經の研究 (Investigations on the Star Manual of Shi Shen; Tokyo, 1929).

[14]Pan Nai, *Zhongguo Hengxing Guance Shi*, 1989, ch. 2, pp. 48 ff.

[15]Yabuuchi Kiyoshi, "To kaigen senkyō chū no seikyō" 唐開元占經中之星經 *Tōhō Gakuhō* (Kyoto), 8 (1937), pp.56-74.

[16]Qian Baocong, "Ganshi xingjing yuanliu kao" 甘石星經源流考, *Guoli zhejiang daxue jikan* 國立浙江大學季刊 (Quarterly Publication of the State University of Zhejiang), 6 (1937).

Y. Maeyama [1977] made a numerical analysis of the data and used a minimum variance estimate to determine the epoch of the star catalogue in *SSXJ*. He found 70 ± 30 BC. He was the first who found a systematic error caused by pole misalignment of the instrument in the catalogue of *SSXJ*[17].

Ueda, Pan Nai and Maeyama based their work on the analysis of the star catalogue in *SSXJ*. However, for analysis of the data, the first step is to identify the star. They all made their own choice for the identification of each leading star of each constellation. Ueda and Maeyama have used the identification made by I. Tsuchihashi. They made some adjustments, however, to get agreement with the data. Pan Nai [1989] referred to the Song sky which he reconstructed according to observations during the Song dynasty and argued that his choice was more reliable as it was closer to ancient times.

3.3 Basic data of the SSXJ *catalogue*

In this section we present basic data concerning the 120 constellations of Shi Shen's school[18] from the *SSXJ* text in *KYZJ* in Table 3.1. To save space we also present the result of our study and necessary comments on these constellations in the same table. This table is the most essential part of this chapter. We will refer to it occasionally.

Because Latitudes (HDD) of the stars do not change in the course of time, they can not be dated by means of precessional calculation. Thus the *huang dao nei wai du* in the SSXJ catalogue are not listed in this table. Only occasionally do we use the HDD to check the QJD data; in that case we will explain.

[17]Y. Maeyama, "The oldest catalogue of China, *Shi shi xingjing*," Prismata-Naturw. Studien, 1977, pp.211-245.

[18]Here we would like to mention that the 28 lunar lodges did not exclusively belong to Shi Shen's school. They existed much earlier than Shi Shen's system of constellations and were the common knowledge of all three schools. But only Shi Shen's school gave the coordinates of the *ju xing* of the 28 lunar lodges.

A simplified armillary sphere made in the Zhengtong period (1437-1442) of the Ming dynasty, now exhibited at the Purple Mountain Observatory in Nanjing. The design was after Guo Shoujing's instrument made during the early Yuan (1280). The meridian circle is essential for observation. Photo taken from the *Zhongguo gudai tianwen wenwu tuji* (Wenwu Press, Beijing; 1980).

Key to Table 3.1, the *Shi shi xingjing* Star Catalogue

I Name of constellation in *pinyin* spelling.

II Number of stars in the constellation.

III Identification of the leading star, the *ju xing* by Yuzuru Ueda [1930].

IV Identification by Yi Shitong [1981].

V Our identification.

VI The distance to the pole(QJD) in Chinese Degrees, presented by Qutan Xida in 729 AD in *Kaiyuan zhanjing* (*SSXJ* data). Han astronomers divided a circle of the celestial sphere into 365.25 degrees corresponding with the number of days in a year. Therefore one Chinese Degree equals 360/365.25 Babylonian Degrees.

VII $(QJD)_{SSXJ}$ - $(QJD)_{id.star}$, both in Babylonian Degrees. The $(QJD)_{id.star}$ is the computered distance to the pole of the corresponding star based on identification at 100 BC (For the 28 *ju xing* at 50 BC).

VIII Notes leading to our comments.

Abbreviations used in the subscripts of Table 3.1

SSXJ *Shi shi xingjing.*

KYZJ *Kaiyuan zhanjing* (729 AD).

QJD *Qu ji du,* the distance to the pole.

RXD *Ru xiu du,* the distance from the *ju xing* of the *xiu* measured in the direction of Right Ascension.

HDD *Huang dao nei wai du,* the distance to the ecliptic which we call Latitude.

Table 3.1: The *SSXJ* Star Catalogue.

No.	I	II	III	IV	V	VI	VII	VIII
1	Sheti	6	η Boo	η Boo	η Boo	59.50	-1.55	
2	Dajiao	1	α Boo	α Boo	α Boo	58.00	-1.70	
3	Genghe	3	ρ Boo	ρ Boo	ρ Boo	48.00	-1.95	note[1]
4	Zhaoyao	1	γ Boo	γ Boo	γ Boo	40.75	-1.24	
5	Xuange	1	λ Boo	λ Boo	λ Boo	32.50	-1.29	
6	Tianqiang	3	κ Boo	κ Boo	κ Boo	28.75	0.83	
7	Tianpei	5	ι Her	ξ Dra	ι Her	42.00	-0.08	note[2]
8	Nüchuang	3	π Her	π Her	π Her	50.00	0.14	
9	Qigong	7	β Boo	δ Boo	β Boo	39.25	-1.40	
10	Guansuo	9	π Ser	π CrB	β CrB	59.25	6.24	note[3]
11	Tianji	9	ξ CrB	ξ CrB	ξ CrB	51.50	-1.78	
12	Zhinü	3	α Lyr	α Lyr	α Lyr	52.00	-0.20	
13	Tianshi	22	ζ Oph	ζ Oph	ζ Oph	94.25	-1.03	
14	Dizuo	1	α Her	α Her	α Her	71.25	-0.81	
15	Hou	1	α Oph	α Oph	α Oph	73.75	-0.94	
16	Huanzhe	4	33 Oph	32 Oph	60 Her	72.50	-0.73	
17	Dou	5	ω Her	ω Her	κ Her	72.00	-3.83	
18	Zongzheng	2	γ Oph	β Oph	γ Oph	84.00	-1.23	note[4]
19	Zongren	4	67 Oph	66 Oph	68 Oph	85.00	-2.33	note[4]
20	Zong	2	71 Oph	110 Her	71 Oph	79.00	-1.26	note[4]

— *continued on next page*

[1]The present *SSXJ* text gives the wrong value 38 instead of 48 for the QJD. Judging by the HDD value also given in the *SSXJ*, 48 is the right value for the QJD. Moreover, the descriptions about *Genghe* in the text of *SSXJ* obviously indicate a group of three stars ϵ, σ, ρ Bootes as *Genghe*; so the value 48 fits the celestial circumstance. Apparently a transcription error has happened here, mistaking the number 3 for the number 4. The same error has also happened to the QJD of the leading stars of *Tianpei* and *Laoren*. They are all 10 degrees too low.

[2]QJD 10 degrees off. See note[1].

[3]*Guansuo* is a chain which is a loop of links. In the sky we clearly see the stars of the Corona Borealis in a brilliant circle. The descriptions in the text of *SSXJ* suggests β CrB as leading star, although the cross-mark on our computed star maps according to the QJD and RXD in the *SSXJ* agrees better with Ueda's choice for the leading star.

[4]One could choose β Oph as well for the leading star of this constellation if only considering the distance of a star to the cross-mark on the computed maps of 100 BC. But the *SSXJ* text described clearly that it was the southern one of the group of two that was the leading star. The southern one, γ Oph, was within reasonable distance to the cross-mark as well. For constellations composed of stars close to each other, we gave priority to the descriptions in the text.

Table 3.1: The *SSXJ* Star Catalogue, *continued*

No.	I	II	III	IV	V	VI	VII	VIII
21	Dongxian Xixian	8	ω Oph	φ Oph	ψ Oph	103.00	-1.19	
22	Tianjiang	4	99G Oph		θ Oph	111.00	-1.30	
23	Jianxing	6	ξ Sgr	ξ Sgr	ξ Sgr	113.25	0.78	note[1]
24	Tianbian	9	1 Aql	α Sct	5 Aql	90.75	-0.86	note[2]
25	Hegu	12	α Aql	α Aql	α Aql	85.00	-0.53	
26	Lizhu	5	1 Aqr	70 Aql	1 Aqr	94.00	-2.00	
27	Hugua	5	ζ Del	α Del	ζ Del	81.50	-0.21	
28	Tianjin	9	δ Cyg	γ Cyg	δ Cyg	49.00	-0.22	note[3]
29	Dengshe	22	α Lac	α Lac	α Lac	51.00	0.70	
30	Wangliang	5	β Cas	β Cas	β Cas	42.50	-0.32	
31	Gedao	6	δ Cas	ι Cas	φ Cas	43.25	-0.72	
32	Fulu	1	γ Cas	ζ Cas	υ Cas	43.00	-0.05	
33	Tianda-jiangjun	11	γ And	γ And	γ And	60.33	0.64	
34	Daling	8		9 Per	9 Per	44.25	-1.51	
35	Tianchuan	9	η Per	η Per	η Per	43.50	-1.70	
36	Juanshi	6		ν Per	ν Per	56.00	-1.18	
37	Wuche,Zu	14	ι Aur	ι Aur	ι Aur	63.00	-0.93	
38	Tianguan	1	ζ Tau	ζ Tau	ζ Tau	73.50	-0.57	

— *continued on next page*

[1] The *SSXJ* text says the leading star for this constellation was the most western one. According to this we chose ξ Sgr as the leading star, although the derived Right Ascension and Declination from the given RXD and QJD values could correspond to two other stars of the constellation as well.

[2] According to the cross-mark on our computed star maps we chose the star 5 Aquilae, now a faint star of the 6th magnitude. It is a variable B-type star on top of the hat, *Tianbian*. Astrophysists know that this type of star might change its visual brightness appreciably in the course of only a few centuries.

[3] *SSXJ* gives the RXD of the leading star of *Tianjin* with respect to *Dou xiu*, suggesting that it is in *Dou xiu*. Obviously it was a transcription error because Tianjin was such a conspicuous constellation in the Celestial River, *Tian He*, the Milky Way, that it could never be misconfigured as extending as far as indicated by the cross-mark. Tianjin means wading place in a river. It certainly should be in the River. As the QJD corresponds best with δ Cyg, it is obvious that its lodge was *Niu xiu*.

Table 3.1: The *SSXJ* Star Catalogue, *continued*

No.	I	II	III	IV	V	VI	VII	VIII
39	Nanbeihe	6	β CMi	ε CMi	β CMi	80.00	-1.33	
40	Wuzhuhou	5	θ Gem	θ Gem	θ Gem	57.00	-0.67	
41	Jishui	1	χ Gem	65 Aur	o Gem	55.00	0.33	
42	Jixing	1	65 Gem	κ Gem	φ Gem	61.50	-0.19	
43	Shuiwei	4	74 Gem	6 CMi	68 Gem	72.50	-1.04	
44	Xuanyuan	17	α Leo	α Leo	α Leo	71.00	0.56	
45	Shaowei	4	53 Leo	52 Leo	53 Leo	70.50	-0.07	note[1]
46	Taiwei	10	β Vir	β Vir	β Vir	76.50	-1.51	
47	Huangdizuo	1		β Leo	β Leo	63.50	-1.61	note[2]
48	Shidizuo	4						note[2]
49	Neiping	4		ξ Vir	ξ Vir	72.50	0.58	note[2]
50	Langwei	15		γ Com	γ Com			note[2]
51	Langjiang	1		31 Com	α CVn	39.25	-1.39	note[2]
52	Changchen	7		α CVn	61 UMa	45.00	-0.26	note[2]
53	Santai	6	18 UMa	ι UMa	ι UMa	36.25	-1.10	note[3]
54	Xiang	1	χ UMa	5 CVn	χ UMa	31.50	-0.20	
55	Taiyangshou	1	ψ UMa	χ UMa	ψ UMa	35.50	-0.33	
56	Tianlao	6	44 UMa	ω UMa	44 UMa	26.50	0.21	
57	Wenchang	6	o UMa	28 UMa	o UMa	25.75	-1.09	

— *continued on next page*

[1]This constellation might indicate completely different stars at the Han time, as compared with star maps after the Han. The text of *SSXJ* said: " The four stars of Shaowei were located west of the western wall of *Taiwei*. They line up from South to North." From this description we guess that these four stars were 53, 52, 51 Leo and 41 LMi. The cross-mark for the leading star is at 53 Leo.

[2]All the constellations from No. 47 to No. 52 concern mythological representations in and about *Taiwei*. The QJD of these six constellations are missing in the *KYZJ*. But we found the QJD values for four of these six constellations in another book from the Tang, the *Tiandi ruixiang zhi* (Records of Omina from Heaven and Earth). The descriptions in *Tianguan shu* and *Jin shu* are so clear that one can identify these stars.

[3]The QJD of *Santai* given by the *KYZJ* was 30.25 Chinese Degrees which would agree with Ueda's identification, 18 UMa. But in the *Tiandi ruixiang zhi* we find 36.25 Chinese Degrees which corresponds better with ι UMa, the clearly visible "higher step" of the Three Steps *Santai*.

Table 3.1: The *SSXJ* Star Catalogue, *continued*

No.	I	II	III	IV	V	VI	VII	VIII
58	Beidou,Fu	8	α UMa	α UMa	α UMa	18.25	-0.62	note[1]
59	Ziwei	15		α Dra	κ Dra	9.50	0.68	note[2]
60	Beiji Gouchen	11	α UMi	α UMi	α UMi	11.50	-0.97	note[3]
61	Tianyi	1	ι Dra	10 Dra	7 Dra	10.50	-1.23	note[4]
62	Taiyi	1			8 Dra	10.00	-1.76	note[4]
63	Kulou	29		ζ Cen	δ Cru	140.00	0.76	
64	Nanmen	2		ε Cen	ξ_2 Cen	130.00	-0.15	
65	Ping	2		γ Hya	γ Hya	100.00	-3.01	
66	Qiguan	27		γ Lup	c_1 Cen	115.50	-0.99	
67	Jizu	12		θ Lup	η Lup	124.50	2.75	
68	Gui	5		ε Ara		131.00	-1.04	
69	Fuyue	1		G Sco	G Sco	120.50	-4.45	
70	Yu	1		M7 Sco	M7 Sco	122.00	-1.20	

— continued on next page

[1]*Beidou* and *Fu* were mentioned together in *SSXJ*, numbering 8 stars. *Fu* (80 UMa) was regarded as the Operator of the *yang*-valve *Kaiyang* (ζ UMa), the sixth star of *Beidou*. *Fu* is the legendary Assistant of Shangdi, regulating growth and decline in Nature.

The *SSXJ* text also gives the coordinates of the "fifth star" of *Beidou*, *Yuheng* (ϵ UMa). They are unrecognizable however because of transcription errors. *Yuheng* is the Jade Observation Tube used to observe the position of stars.

[2]*Ziwei*, consisting of 15 stars can only mean *Ziwei Yuan*, the walls of *Ziwei*. The position of the leading star according to the coordinates in the *SSXJ* can only mean κ Dra. The *SSXJ* text says " the leading star is the right pivot of the gate of *Ziwei*." This suggests that α Dra in this case is not the right pivot but the left pivot, although all star maps after the Han indicated the opposite. Apparently the wall of *Ziwei* was designed in a different way during the Han.

[3]*Beiji* and *Gouchen* were indicated together in the *SSXJ* as a conglomerate of 11 stars. Apparently there was still a lot of discussion during Sima Qian's lifetime about the internal organization of *Ziwei*. Only the prominent α UMa was indicated as the leading star.

[4]*Tianyi* and *Taiyi* were faint stars deliberately chosen to indicate their holy and mysterious character as August Unity. According to Sima Qian *Tianyi* was also called *Yinde*, "existing of three stars between *Beiji* and *Beidou*, hardly visible, in the shape of a cone, with its peak pointing to the North." By looking at the maps we see these three stars are 7, 8, 9 Dra. *Taiyi* has almost the same meaning as *Tianyi*. Probably these three stars were an indication of a Trinity in the August Unity in Heaven.

Table 3.1: The *SSXJ* Star Catalogue, *continued*

No.	I	II	III	IV	V	VI	VII	VIII
71	Chu	3		σ Ara		132.50	-0.36	
72	Bie	14		α Tel	μ CrA	129.50	-0.76	
73	Jiukan	9			ι Sgr	136.00	0.11	
74	Baijiu	4		γ Gru	ι PsA	131.25	-1.02	
75	Yulin	57		29 Aqr	υ Aqr	120.75	-0.97	
76	Beiluo	1		α PsA	α PsA	130.75	-0.61	
77	Tusikong	1		β Cet	β Cet	120.25	-1.11	
78	Tiancang	6		ι Cet	η Cet	112.00	-1.33	note[1]
79	Tianqun	13		α Cet	α Cet	96.50	-0.62	
80	Tianlin	4		5 Tau	o Tau	90.00	-1.46	
81	Tianyuan	16		γ Eri	γ Eri	114.00	1.02	
82	Shenqi	9		o_1 Ori	π_5 Ori	93.00	-1.70	
83	Yujing	4		λ Eri	β Eri	100.75	-0.81	note[2]
84	Ping	2		μ Lep	ε Lep	118.00	-0.90	
85	Ce	4		α Lep	β Lep	115.00	-1.18	
86	Tianshi	1		μ Col	ν_2 Col	123.00	-0.65	
87	Junshi	13		β CMa	17 Lep	110.00	0.17	
88	Yeji	1		ν_2 CMa	β CMa	111.00	0.26	
89	Lang	1		α CMa	α CMa	106.75	-0.83	
90	Hu	9		δ CMa	κ CMa	122.75	-0.98	
91	Laoren	1		α Car	α Car	143.50	-1.23	note[3]
92	Ji	5		γ Vel	δ Vel	138.00	-2.12	note[4]

— *continued on next page*

[1]The original *SSXJ* text in the *KYZJ* gave a QJD of 120 Chinese Degrees. But *Tiandi ruixiang zhi* gave 112 Degrees, which is better.

[2]Large error in QJD value; 20 degrees off.

[3]See note for No.3. Large error in QJD value; 10 degrees off.

[4]The original *SSXJ* text in *KYZJ* gave a QJD of 148 Chinese Degrees, but *Tiandi ruixiang zhi* gave the value 138.

Table 3.1: The *SSXJ* Star Catalogue, *continued*

No.	I	II	III	IV	V	VI	VII	VIII
93	Jiao 角	2	α Vir	α Vir	α Vir	91.00	-0.28	12[1]
94	Kang 亢	4	κ Vir	κ Vir	κ Vir	89.00	-1.98	9
95	Di 氐	4	α Lib	α Lib	α Lib	94.00	-3.48	16
96	Fang 房	4	δ Sco	π Sco	π Sco	108.00	-1.70	5
97	Xin 心	3	σ Sco	σ Sco	σ Sco	108.50	-1.47	5
98	Wei 尾	9	μ_1 Sco	μ Sco	μ Sco	120.00	-3.59	18
99	Ji 箕	4	γ Sgr	γ Sgr	γ Sgr	118.00	-1.25	11
100	Dou 斗	6		φ Sgr	φ Sgr	116.00	-1.83	26.25
101	Niu 牛	6	β Cap	β Cap	β Cap	110.00	-0.37	8
102	Nü 女	4	ε Aqr	ε Aqr	ε Aqr	106.00	-0.32	12
103	Xu 虛	2	β Aqr	β Aqr	β Aqr	104.00	-0.21	10
104	Wei 危	3	α Aqr	α Aqr	α Aqr	99.00	-1.18	17
105	Shi 室	2	α Peg	α Peg	α Peg	85.00	-1.23	16
106	Bi 壁	2	γ Peg	γ Peg	γ Peg	86.00	-1.20	9
107	Kui 奎	16	δ And	ζ And	ζ And	77.00	-1.98	16
108	Lou 婁	3	β Ari	β Ari	β Ari	80.00	-1.30	12
109	Wei 胃	3	35 Ari	35 Ari	35 Ari	72.00	-1.43	14
110	Mao 昴	17	7 Tau	17 Tau	17 Tau	74.00	-1.40	11
111	Bi 畢	8	ε Tau	ε Tau	ε Tau	78.00	-0.85	16
112	Zhi 觜	3	λ Ori	λ Ori	φ Ori	84.00	-1.16	2
113	Shen 參	10	δ Ori	ζ Ori	δ Ori	94.50	-1.01	9
114	Jing 井	8	μ Gem	μ Gem	μ Gem	70.00	-0.40	33
115	Gui 鬼	5	δ Cnc	θ Cnc	θ Cnc	68.00	-0.45	4
116	Liu 柳	8		δ Hya	δ Hya	79.00	-1.52	15
117	Xing 星	7		α Hya	α Hya	91.00	-1.83	7
118	Zhang 張	6		υ Hya	υ Hya	97.00	-1.20	18
119	Yi 翼	22	ν Hya	α Crt	α Crt	99.00	-0.43	18
120	Zhen 軫	4	γ CrV	γ CrV	γ CrV	99.00	1.29	17

[1]This page of Table 3.1 gives basic data for the twenty eight lunar lodges (*xiu*). Because the *pinyin* spellings of some *xiu*-names are the same and thus are confusing, we add the corresponding Chinese characters to all *xiu*-names to distinguish between them. The width *ju du* of each *xiu* is also given in columnn VIII for completeness of the data of the twenty eight *xiu*.

3.4 *Preliminary analysis based on the 28* xiu

Of the above three kinds of coordinates, only the QJD is sensitive to time. So it could be used to determine the observation time of the star catalogue. We will first use the QJD of the determinative stars of the 28 *xiu* as a sample of data for this purpose. We chose this special set of data to determine the time of observation because:

1. these determinative stars were standard reference points in the sky for the coordinate system. They must have been measured with great care. These data can be expected to be more reliable than those of other stars.
2. the determinative stars of the *xiu* have been well followed by other ancient astronomers. Not many mistakes are to be expected in the identification of these stars. Scholars agree quite well about the corresponding "standard names" of these stars.
3. they are systematically distributed all around the equatorial belt which makes them statistically attractive for analysis.
4. In *KYZJ* the QJD values for the 28 *xiu* appear in the main text while those for other constellations appear in small-printed notes between the data lines of the main text. That means the group of the 28 *xiu* was more important than the group of the other 92 constellations and they were treated differently by the author of *KYZJ*.

3.4.1 Instruments and the method of observation

From the text in *KYZJ* we can also deduce the method of observation, knowing that the determinative star was almost always the most western star in the constellation. The text carries the two characters "*xian zhi* 先至" before each QJD of a determinative star. "*xian zhi*" means "coming first" or "passing first". This suggests meridian observation. Considering the old tradition of the observation of star-culminations we get the impression that the instrument was probably an armillary sphere, at least a primitive type with a vertical circle as meridian.

The armillary sphere was invented at the end of the Former Han when the *hun tian* theory of cosmology was introduced. The recognition of the celestial sphere was essential in this theory. Before the Han astronomical measurements were done by means of the gnomon and water clock[19]. The width of the *xiu, ju du* 距度 could be measured but not the distance to the pole[20]. An armillary sphere, or strictly speaking an instrument incorporating circles, was necessary for the measurement of the QJD. In the second year of the reign of Ganlu 甘露 (52 BC), Geng Shouchang already used an instrument called *yuan yi* 圓儀 (Circular Instrument) to do measurements[21]. Judging by the name of the instrument we know it might be an example of preliminary types of an armillary sphere.

The basic data of the *ju xing* of the *xiu* are listed in Table 3.1 from No.93 to No.120.

3.4.2 Systematic errors

When one wants to analyze a set of astronomical observations the first thing to do is to investigate the possibility of a systematic error in the observations. In case of meridian observations, a misalignment of the pole of the instrument can be a trivial reason for such a systematic error in all QJD. See Figure 3.4.

$$C = P_i P_i' = -\delta_1. \tag{3.1}$$

Here we denote δ_1, with negative sign, as a systematic error in all Declinations caused by the misalignment of the pole of the instrument.

A second reason for systematic errors in the Declinations derived from the measured QJD of the determinative stars is the precessional motion of

[19]According to the *Zhou li* 周禮 (Rituals of the Zhou), there were special officials in charge of the gnomon and water clock: *yu ren* 玉人 (the official in charge of the Jade Gnomon) and *qi hu shi* 挈壺氏 (the official in charge of the clepsydras).

[20]*Zhoubi suanjing* 周髀算經 introduced a way to measure the *ju du* only by gnomon. We doubt the feasibility of this. The usual way to measure the *ju du* was to record the time difference between stars passing through the meridian which was marked by a vertical stick to the observer's south.

[21]See the *Xu han shu lüli zhi* (Treatise on Pitchpipes and Calendar of the Continued History of the Han). The *Xu han shu* 續漢書 was written by Sima Biao 司馬彪 (c. 240 - c. 306). But most parts of the book were missing, only the treatises of the book were preserved till today because Liu Zhao 劉昭 (fl. 502 - 520) incorporated it into the *Hou han shu* 後漢書 (History of Later Han) by Fan Ye 范曄 (398 - 446). Thus this treatise is also referred to as the *Hou han shu lüli zhi*.

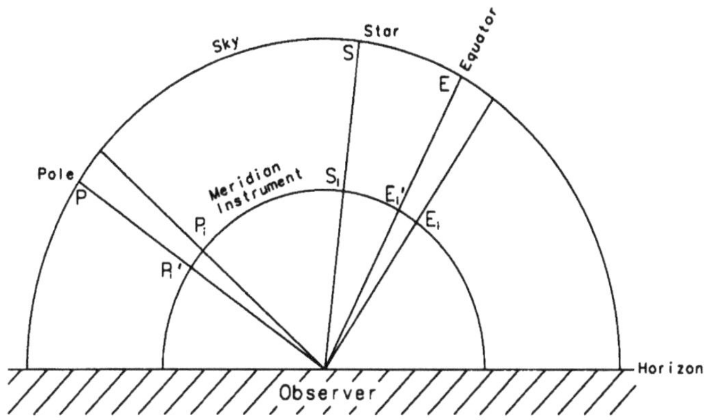

Figure 3.4: The misalignment between the celestial pole and the instrumental pole. The distance of the star to the pole (QJD) as measured on the meridian instrument is S_iP_i and the derived Declination E_iS_i. In reality the celestial coordinates are SP and ES, respectively.

the pole and of the celestial equator. Actually one can use the precession to determine the epoch of the declinations if one can find the equatorial plane of the time which fits best to this set of data.

By using basic data of stars in the Bright Star Catalogue [1964] and the rigorous precession formulae from the Astronomical Almanac [1990] we calculate the Declination D_s for each star at a supposed time t_s. If the observations were done in reality at the time t_r, then the "systematic" error δ_2 is given by $D_s - D_r$

$$\delta_2 = D_s - D_r = n(t_s - t_r)\cos\alpha \tag{3.2}$$

where $n = 0.5567$ degrees per 100 Years, which is the precession constant; α is the Right Ascension of the star at the supposed time of observation t_s; $t_s - t_r$ is the time difference between a supposed time of observation and the real time of observation.

For simplicity we write $n(t_s - t_r) = A$, so the total systematic error is

$$\delta = \delta_1 + \delta_2 = C + A\cos\alpha. \tag{3.3}$$

Naturally the supposed time for which A equals zero is the real time of observation.

To find the real time of observation we will scan a time-span of a few centuries around the epoch of one year BC by assuming several dates t_s between 500 BC and 450 AD. To find C and A in equation 3.3 we apply Fourier analysis within this time span on the total set of original data from *SSXJ*, as presented in the 6th column of Table 3.1.

3.4.3 Fourier analysis of the data

Each of the Declinations D_0 from the 6th column of Table 3.1 differs from the calculated Declination D_s by

$$\Delta\delta = D_s - D_0 \tag{3.4}$$

from which we will derive a systematic error in the form of equation 3.3. The total set of 28 $\Delta\delta$ values provides us with a zigzag function $f(\alpha)$ with the Right Ascension α as a variable.

The Fourier cosine transform of $f(\alpha)$ can be written as

$$f(\alpha) = \sum_{i=1}^{\infty} b_i \cos(i-1)\alpha \tag{3.5}$$

from which follows

$$b_1 = 1/2\pi \int_0^{2\pi} f(\alpha)d\alpha = C \tag{3.6}$$

$$b_2 = 1/\pi \int_0^{2\pi} f(\alpha)\cos(\alpha)d\alpha = A. \tag{3.7}$$

The values of C and A are listed in Table 3.2. For every 50 years from 500 BC to 450 AD we produced a graph of the zigzag line and the sum of the first two terms of the Fourier expansion as shown in Figure 3.5.

3.4.4 Estimate of the observation time of the *SSXJ* star catalogue

The result of the Fourier analysis is that C almost always equals -1.02, suggesting that the polar axis of the meridian instrument with which the observations have been made had a misalignment of 1.02 degrees relative to the

Year	$C(^o)$	$A(^o)$	SRV(^{o})	Year	$C(^o)$	$A(^o)$	SRV(^{o})
-500.00	-1.03	-2.48	0.56	0.00	-1.02	0.28	0.11
-450.00	-1.03	-2.21	0.50	50.00	-1.02	0.56	0.15
-400.00	-1.03	-1.93	0.44	100.00	-1.02	0.83	0.20
-350.00	-1.03	-1.66	0.38	150.00	-1.02	1.11	0.25
-300.00	-1.03	-1.38	0.32	200.00	-1.02	1.38	0.30
-250.00	-1.03	-1.11	0.27	250.00	-1.02	1.66	0.35
-200.00	-1.03	-0.83	0.21	300.00	-1.02	1.93	0.40
-150.00	-1.02	-0.55	0.16	350.00	-1.02	2.21	0.46
-100.00	-1.02	-0.27	0.12	400.00	-1.02	2.48	0.51
-50.00	-1.02	0.00	0.10	450.00	-1.02	2.75	0.56

Table 3.2: List of C and A values after the Fourier analysis of 25 *ju xing* for 20 supposed times of observation, every 50 years from $t_s = -500$ years to $t_s = +450$ years. We did not use the *ju xing* of *xiu Di* 氐 (α Lib), *Wei* 尾 (μ Sco) and *Zhen* 軫 (γ Crv) because of obvious errors in the QJD of these *ju xing*. See Figure 3.6. The fourth column gives SRV, the square root variance. All six columns are in Babylonian degrees.

direction of the axis of the earth, i.e. to the celestial pole. Moreover we see $A = n(t_s - t_r)$ is at a minimum of nearly zero degrees for 50 BC. That means the observation is estimated to have been made at this time. The real systematic error of the observation is due to the deviation of the polar axis of the instrument.

After elimination of the systematic error $C+A\cos(\alpha)$ we also calculated the square root variance SRV of the observations, as listed in Table 3.2.

$$SRV = \frac{1}{28}\sqrt{\sum_{i=1}^{28} \Delta_i^2} \tag{3.8}$$

while for each *ju xing*

$$\Delta_i = (\Delta\delta - (C + A\cos(\alpha)))_i. \tag{3.9}$$

Detailed calculation showed that A equals zero at 60 years BC. whereas SRV equals 0.17 degrees. If we remove three *ju xing*[22] with obvious large

[22]These three *ju xing* are of *xiu Di* (α Lib), *Wei* (μ Sco) and *Zhen* (γ Crv). See Figure 3.6.

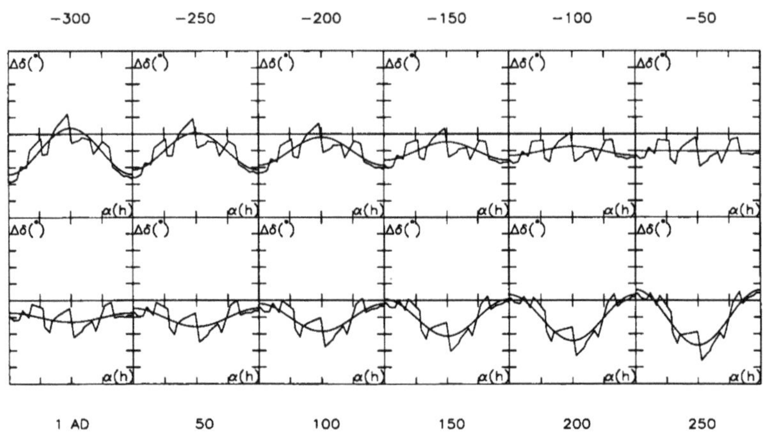

Figure 3.5: Analysis of errors in Declination based on the Fourier cosine transform for 25 *ju xing*. The zigzag line represents the empirical spread of the points $\Delta\delta$ for 12 supposed dates; the cosine transform line represents the systematic deviation δ according to equation 3.3.

errors (See Table 3.1) we obtain 50 BC for A and 0.10 degrees for SRV. Therefore we estimate the uncertainty of our method of age determination to be

$$\Delta t = (0.10/n) \times 100 = (0.10/0.556) \times 100 \simeq 18 years. \qquad (3.10)$$

From the above analysis we estimate that the *ju xing* of the *Shishi xingjing* star catalogue were observed around 50 BC.

We show in detail the results of our analysis for 28 *ju xing* and for 25 *ju xing* in Figure 3.6. In the latter case we left the *ju xing* for *Di, Wei* and *Zhen* out. The zigzag curve represents the empirical spread of $\Delta\delta$. See equation 3.4. The line at a distance of 1.02 degrees below zero represents the cosine transform line of the errors. For $t_s = -50$ years this is practically a straight line. The scatter in the lower curve demonstrates the accuracy of the measurements of the QJD of the *ju xing* which were possibly done by Xianyu Wanren's group. See Section 3.5.3.

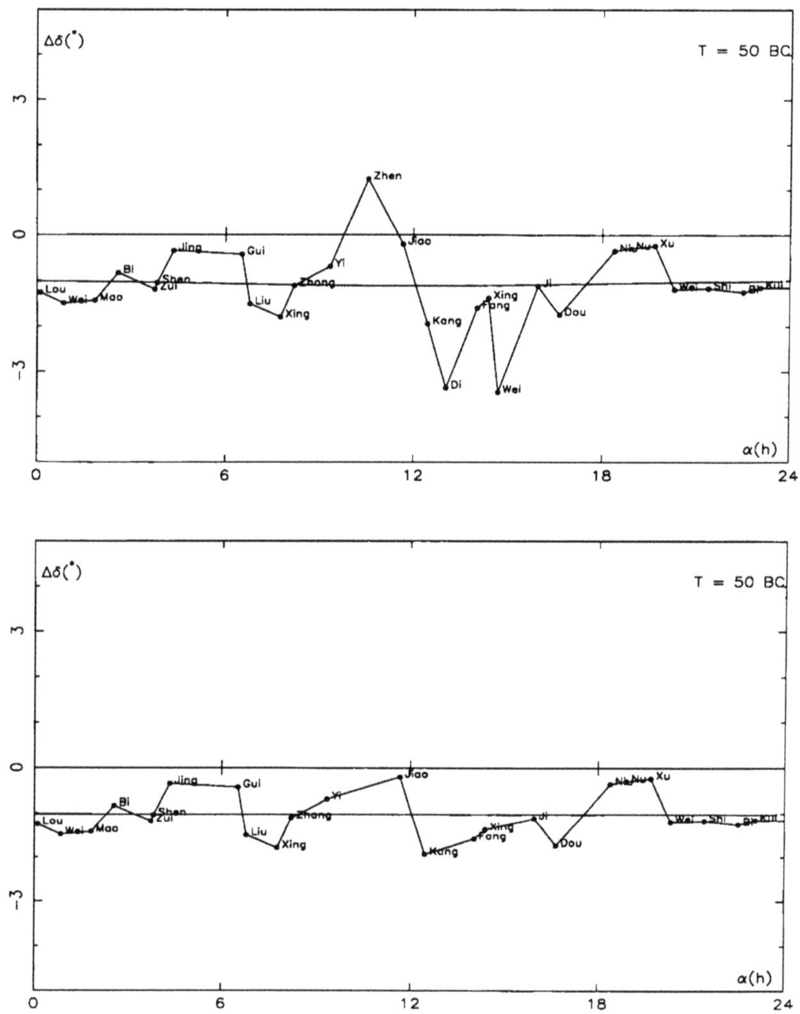

Figure 3.6: Spread in Declination for the epoch 50 BC after the Fourier analysis of 28 *ju xing* (above) and of 25 *ju xing* (below).

3.5 Analysis of the whole star catalogue

3.5.1 The problem of identification in the Chinese sky

Earlier researchers have come to different conclusions about the observation time of the *SSXJ* catalogue. One essential reason has been that they have used different identifications of the leading stars. An immediate and essential question in studying the ancient Chinese sky is: Which stars did ancient astronomers actually refer to when naming and modelling a constellation? When the Jesuits and some Chinese scholars made use of the European sky to compile new star catalogues of the Chinese sky, they were confronted with the same question and had to apply their own solution. But their choices of the corresponding stars were not supported by their own work[23]. Thus the identification problem was left to historians.

The first important effort at identification was done by the Dutch scholar Schlegel in 1875. In his book *Uranographie Chinoise* he verified 759 names of constellations and single stars and made a correlative star map with Chinese and Western star names. For historical verification he used the star map of Xu Fa 徐發 (early Qing) in his book *Tianyuan lili quanshu* 天元曆理全書 (Complete Treatise on Calendars (1682)). For correlation of stars he used Argelander's star map which was published in 1840 in Germany.

Most later identification works have been based on *Yixiang kaocheng* (Imperial Astronomical Instruments) compiled by the German Jesuit I. Kögler (1680 - 1746) together with some Chinese astronomers. *Yixiang kaocheng* contains a star catalogue which was essentially a transformation of Flamsteed's star catalogue, probably supplemented with some observations of their own.

In 1912 I. Tsuchihashi published the *"Catalogue d'etoiles observées a Pekin sous l'empereur K'ien Long"* in which there were four pages with correlative star maps of the Chinese sky. This work was based on the *Yixiang kaocheng* and was widely used by scholars.

The most recent identification work was done by Yi Shitong [1981]. He produced a bilingual star atlas with 22 maps, together with a catalogue containing the Chinese star names and their corresponding modern standard names. His work was also based on the *Yixiang kaocheng*. Identification of

[23]See Chapter 2.

stars was done by referring to the Boss [1936] star atlas.

But the real identification question remained unsolved because all these works were based on the work of the Jesuits who sometimes simply transferred a European star catalogue into a Chinese one and their identification with the Chinese constellations was based on Chinese catalogues and star maps from not earlier than the Song dynasty (1050 AD). They had no data about the constellations of the Han.

In the present study we will make our own identifications of the leading stars from the *SSXJ* star catalogue.

3.5.2 Star map of 100 BC and the identification of the leading stars

In principle the above Fourier analysis could also be applied to the other 92 stars and constellations presented by *Shi Shi*. However the problem is that the identification of the leading stars of these constellations is not as certain as those of the 28 *xiu*. Sometimes later investigators disagreed about the choice of a leading star. There are several reasons that identification is difficult:

— the data of these stars are less accurate than of the 28 *ju xing*;

— in extended constellations the choice of a leading star is more or less arbitrary;

— historical and philosophical arguments were used to change the shape of some constellations which affected the choice of the most western star;

— precession caused big changes in visibility along the southern horizon, but the memory of what had been visible remained and influenced later presentations.

All former identifications of the Chinese sky were based on material from not earlier than the Song dynasty. The correspondence with the *SSXJ* was not certain. It could not be assumed to be a representation of Shi Shen's school.

As we have seen above, the main activities of Shi Shen's school were during the Former Han. In books before Sima Qian's *Shi ji* only Shi Shen's name was mentioned. Astronomical information was usually very vague. It was only after Sima Qian that there were frequent "quotations" of Shi Shen. But these "quotations" contained astronomical knowledge which could only be related with Han philosophy and technology. The QJD of the leading stars of the constellations of Shi Shen's school were apparently observed during the Han.

Making calendars was one of the most important purposes of astronomical measurements in ancient China. But for this purpose the measurement of the *qu ji du* 去極度 (Distance to the Pole) was not definitely necessary. The measurements of the *ju du* 距度 (Width of the *xiu*) and of the *ru xiu du* 入宿度 (Degrees into the *xiu*) were enough for the basic calendrical calculations in the ancient *si fen* 四分 calendar. Only the difference in Right Ascensions was of importance. The measurement could be done with the gnomon and the water clock. It seems that before the Han there was no interest in the measurement of the QJD of stars. Only when people became interested in star maps or celestial globes to illustrate the whole sky did it become necessary to measure the QJD with a meridian instrument or armillary sphere. This modernization was stimulated by the important Calendar Reform of 104 BC, during the Former Han.

From the astronomical chapter of the *Han shu* 漢書 (History of the Former Han Dynasty), we know that Xianyu Wanren 鮮于妄人 organized a group of 22 astronomers to do observations to verify the validity of the new calendar about 30 years after the Reform. Typical for the astronomical activities of those days are the remarks made by the philosopher Yang Xiong 揚雄 (53 BC - AD 18) who mentioned in his book *Fa yan* 法言 (Admonitory Sayings):

> If somebody asks me about the theory of *hun tian* 渾天 (Enveloping Sky) and about the armillary sphere, I would say Luoxia Hong 落下閎 constructed it, Xianyu Wanren 鮮于妄人 used it for observation and Geng Shouchang 耿壽昌 illustrated it with a celestial globe.

We know that all three persons were active between the Calendar Reform and the end of the Western Han.

From these facts and our analysis based on the 28 *xiu* it seems logical to assume that the coordinates of the other 92 leading stars of *SSXJ* were measured on a meridian instrument during the last century BC. Therefore we reproduced the real celestial background, a star map for the epoch of 100 BC for our identification purpose. This star map was made by means of a computer program based on *The Yale Catalogue of Bright Stars.* In the computation we used the rigorous formulae for precession and proper motion from *The Astronomical Almanac of Washington.* On this map we marked the position of the leading stars of the constellations of Shi Shen's school with crosses according to our calculation of the QJD and RXD from the text of *SSXJ*[24].

In this way we could compare with the sky as observed by Han astronomers in Chang'an at 34° N. Lat. This seems to be the most instructive and reliable way to identify the corresponding stars.

For identification we used the following criteria:

1. The star should be bright enough, well visible to the naked eye, so that it could be observed by ancient astronomers. Preferably the star should be brighter than the 5th magnitude.
2. The star to be chosen for identification should be within a reasonable distance to the cross-mark on our computed star maps.
3. The textual description should be considered. This was decisive in several cases.
4. The later star maps in Chinese history should be considered. The star to be chosen should be in the traditional configuration of a constellation. Exceptions need careful consideration.
5. Cases in which we supposed that there was a large error in the transcription of the QJD data should be checked against the available RXD and HDD, or against other sources if available.

In this way we could identify 90 leading stars[25], plus 28 *ju xing* for the lunar lodges. They are listed in column V of Table 3.1, together with the choices of Ueda [1930] and of Yi Shitong [1981] for comparison. In the last column we give the notes leading to our comments.

---

[24]The *ru xiu du* was always measured relative to the *ju xing* of the *xiu* of reference.

[25]In *SSXJ* the QJD for 6 constellations were lost. But the QJD for four of the six constellations can be found in another book from the Tang, the *Tiandi ruixiang zhi.*

Year	$C(^o)$	$A(^o)$	SRV(^{o})	Year	$C(^o)$	$A(^o)$	SRV(^{o})
-500.00	-0.82	-2.36	0.21	0.00	-0.86	0.42	0.11
-450.00	-0.82	-2.10	0.19	50.00	-0.87	0.70	0.12
-400.00	-0.83	-1.80	0.17	100.00	-0.86	0.95	0.13
-350.00	-0.82	-1.55	0.16	150.00	-0.90	1.28	0.15
-300.00	-0.82	-1.28	0.14	200.00	-0.90	1.55	0.16
-250.00	-0.83	-1.00	0.13	250.00	-0.90	1.83	0.18
-200.00	-0.83	-0.70	0.12	300.00	-0.89	2.10	0.20
-150.00	-0.83	-0.40	0.11	350.00	-0.89	2.37	0.22
-100.00	-0.84	-0.12	0.10	400.00	-0.89	2.65	0.24
-50.00	-0.86	0.15	0.10	450.00	-0.89	2.93	0.26

Table 3.3: List of C and A values from the Fourier analysis for the whole *SSXJ* star catalogue, every 50 years for a supposed time t_s between -500 years and 450 years. The fourth column gives SRV, the square root variance.

3.5.3 The observation time of the *SSXJ* star catalogue

As a further step we applied Fourier analysis to the data of the whole star catalogue, as we have done with the 28 *ju xing* of the *xiu*.

The results are presented in Table 3.3 and in Figure 3.7.

The observation time of the whole SSXJ star catalogue is

$$78 \pm 20 years BC. \tag{3.11}$$

The analysis also shows that the polar-axis of the instrument was about one degree (0.86) misaligned with the celestial north pole.

The year 78 BC[26] is exactly the same year when Xianyu Wanren's group of 22 astronomers made large scale observations to check the reliability of the new *Tai chu* 太初 calendar produced during the calendar reform of 104 BC.

Historically, the great calendar reform started in the year 104 BC when Sima Qian organized a group of astronomers to do astronomical observations. Among them was Luoxia Hong who invented the armillary sphere. The *SSXJ* catalogue must have been observed after Sima Qian's group started

[26]The exact correspondence between the year 78 of (3.11) and the year of Xianyu Wanren's observations is purely accidental.

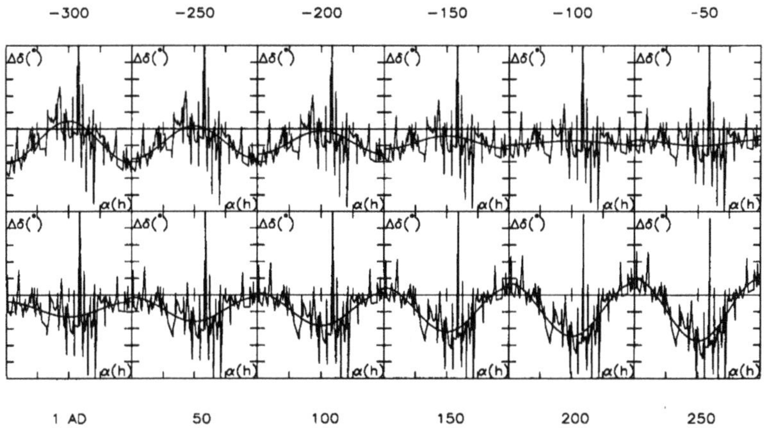

Figure 3.7: Error analysis based on the Fourier cosine transform for the whole *SSXJ* star catalogue. Explanations are given in the caption to Figure 3.5.

to work. Yang Xiong emphasized that it was Xianyu Wanren who made the observations by means of the armillary sphere. According to *Han shu* this happened during the years from 78 to 76 BC, which agrees with our overall age determination.

3.5.4 Remarks on the studies by earlier researchers

The estimated observation times of the *SSXJ* star catalogue by Yabuuchi [1937] and by Qian Baocong [1937] are approximately correct. Our Fourier analysis give almost the same date as Maeyama [1977] has given by his minimum variance method. This proves that his mehtod is original and powerful and certainly strengthens the conclusion that the *SSXJ* star catalogue was observed during one period: between 80 and 70 BC.

The opinion that the *SSXJ* catalogue has been the result of observations during two separate periods (see 3.2.2) does not seem to be correct. Both authors (Ueda Yuzuru and Pan Nai) first assumed that the QJD in the *SSXJ* catalogue were absolutely precise. A date could be found for each QJD value by comparing it with the calculated QJD value of the identified star by means of the precession formulae. From all dates for all leading stars they found two average dates: for Ueda [1933] they were 360 BC and 200 AD and for Pan Nai [1989] they were 450 BC and 160 AD.

We would like to point out that this analysis is incorrect. As we have seen in formula 3.2 the changing of Declinations of stars is not only dependent on the time of observation but also on the Right Ascensions of the stars themselves. That means that each star has its own dependence on time. In this case, the average of all the dates would not be the real observation time of all the stars.

If we look at these two groups of *ju xing*, see Table 3.4, then we see that one is completely on the hemisphere of the spring equinox and the other one on the hemisphere of the autumn equinox. That means for the first group, consisting of the *ju xing* from *xiu Niu* 牛 to *xiu Jing* 井, that the Right Ascensions of these *ju xing* range between -90 an +90 degrees; whereas for the second group, consisting of the *ju xing* from *xiu Gui* 鬼 to *xiu Dou* 斗, the Right Ascensions go between +90 and +270 degrees. This is obviously systematic because according to equation 3.2 one degree systematic error in the QJD would cause the dates of observation of the first group of stars to be about 300 years later on average and for the second group 300 years earlier. Apparently that is the origin of the two epochs.

After adjusting the QJD data by adding a correction of one degree, we calculated the dates for all the *ju xing* by the same method of Ueda and Pan Nai. See Table 3.4. We found that the difference between the two sets of observations before and after Christ disappears. The average of all data revealed a date of about 120 BC, not far from our conclusion. Furthermore, from Figure 3.6 we see that our result based on Fourier analysis fits well with both groups of *ju xing* of Pan Nai. That means there can only have been one time of observation for all the 28 *ju xing*.

Our method of analysis has the advantage that it is based on a real discussion of ancient measurements with respect to possible systematic errors. The influence of misalignment of the pole of the meridian instrument is apparent. Moreover, it is clear that periodic functions like the systematic error in the form of equation 3.3 should always be analyzed on the basis of a Fourier cosine transform which follows the periodic trends in the data in a natural way. The final result is not seriously affected by a few outstanding deviations like those for the determinative stars of *Di* 氐 (α Lib), *Wei* 尾 (μ Sco) and *Zhen* 軫 (γ Crv), as we see from our analysis.

A Lunar lodges (xiu) and id. of *ju xing* by Pan Nai [1989]	B Calculated R. Asc. of *ju xing* for 100 BC (hours)	C Dates in years by Pan Nai	D Dates in years after subtraction of one degree in QJD
Jiao(α Vir)	11.62	-88	+92
Kang(κ Vir)	12.40	-401	-220
Di(α Lib)	12.99	-731	-480
Fang(δ Sco)	13.97	-371	-168
Xing(σ Sco)	14.33	-346	-130
Wei(μ Sco)	14.63	-801	-580
Ji(γ Sgr)	15.90	-406	-76
Dou(φ Sgr)	16.58	-762	-350
Niu(β Cap)	18.34	+359	–
Nu(ϵ Aqr)	18.86	+151	–
Xu(β Aqr)	19.64	+46	-430
Wei(α Aqr)	20.27	+307	-45
Shi(α Peg)	21.35	+188	-45
Bi(γ Peg)	22.46	+188	-45
Kui(ζ And)	23.01	+164	-45
Lou(β Ari)	0.07	+175	-5
Wei(35 Ari)	0.79	+177	+25
Mao(17 Tau)	1.77	+158	+25
Bi(ϵ Tau)	2.51	+60	-105
Zui(φ Ori)	3.70	+461	-40
Shen(δ Ori)	3.78	+318	-55
Jing(μ Gem)	4.29	+92	-330
Gui(θ Cnc)	6.47	-500	+500
Liu(δ Hya)	6.74	–	-625
Xing(α Hya)	7.73	-619	-370
Zhang(υ Hya)	8.17	-436	-75
Yi(α Crt)	9.31	-215	+35
Zhen(γ CrV)	10.51	+186	+375

Table 3.4: Discussion on the two-epoch conclusion about the age of the *Shishi xingjing* star catalogue, taking Pan Nai's analysis of the 28 *ju xing* of *xiu* as example. Column B presents the Right Ascensions of the identified *ju xing* at 100 BC calculated by us using the Almanac [1990] and Bright Star Catalogue [1964]. Column D lists the corresponding calculated dates for the *ju xing* after subtraction of the systematic error of one degree in QJD, computed by us using Pan Nai's method. These rough data would lead to an average date of 120 BC.

3.6 *The sky of* Shi Shi *during the Han*

There were about 783 individual stars mentioned in the 120 constellations of Shi Shen's school. These stars include most bright stars visible on the northern hemisphere. Number of stars , shapes and positions of stars and constellations were described in the *SSXJ*.

In the *Tianguan shu* Sima Qian gave a systematic description of the celestial position of the constellations. He even gave the number of stars in many constellations. His system was essentially the same as that presented in *SSXJ*.

Moreover, in the *Han shu* (History of the Former Han) there is an astronomical chapter in which it is stated that

> There are up to 780 bright stars which are grouped in 118 constellations.

We are therefore certain that the constellations of Shi Shen's school presented the main aspects of the Chinese sky as seen during the Han.

Now we are ready to reconstruct the constellations of Shi Shen's school, as presented in the star maps included in this book. In doing so we took account of the description of the constellations in the text of *SSXJ* as well as of the description by Sima Qian in his *Tianguan shu*. We also compared our results with the star maps of the Song dynasty as well as with all other later identification studies[27]. Although some changes have occurred in the course of history, the main aspects of the Chinese sky have been preserved exceedingly well. The simple reason is that already during the Han there existed such an excellent star catalogue as the *SSXJ* star catalogue. It fixed and confirmed the tradition of the ancient Chinese sky in a satisfactory way.

Full knowledge of this fact is important for the study of the cultural backgrounds of the Chinese civilization as it was projected in the sky. The additions made by Gan De's and Wuxian's Schools apparently were supplements to illustrate in detail various aspects of ancient Chinese culture according to the philosophical and astro-mythological views of the Han time.

---

[27]There exist two star maps from the Song. One is the *Suzhou Tianwen Tu* from 1193, made by Huang Shang. Another is the set of star maps of Su Song in his book *Xin yixiang fayao* of 1094. Yi Shitong's star atlas of 1981 has been our most valuable reference book.

On this basis it is now possible to study the significance of Chinese star names and constellations and compare them with other Asiatic civilizations.

We present the sky of *Shi Shi* on a computed celestial background at the date of 100 BC. The positions of the leading stars according to the *SSXJ* star catalogue were marked with crosses, as one can see on the following maps.

To assist Western readers to find their way on Chinese star maps the abbreviated names of Ptolemaean constellations, plus standard Bayer-Flamsteed star numerations have been inserted.

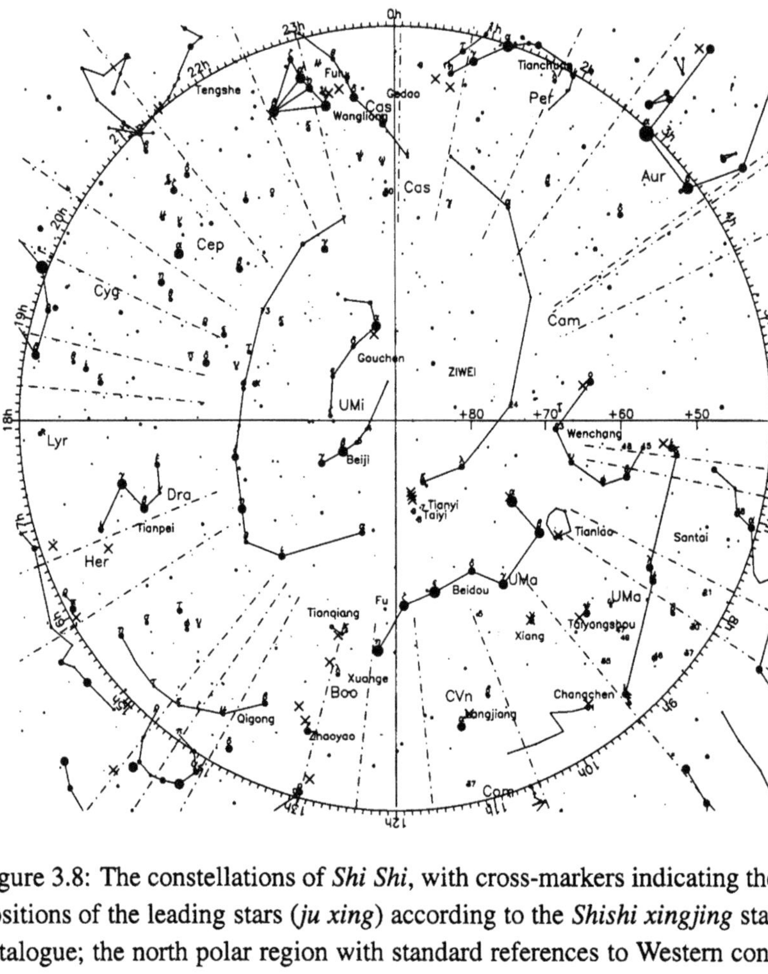

Figure 3.8: The constellations of *Shi Shi*, with cross-markers indicating the positions of the leading stars (*ju xing*) according to the *Shishi xingjing* star catalogue; the north polar region with standard references to Western constellations and with standard star-numeration.

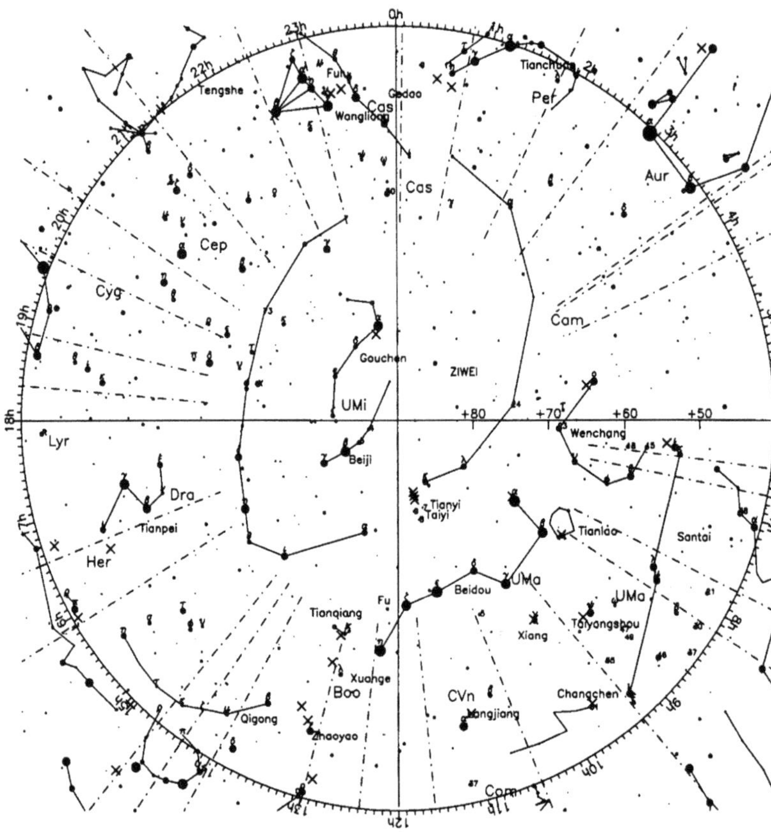

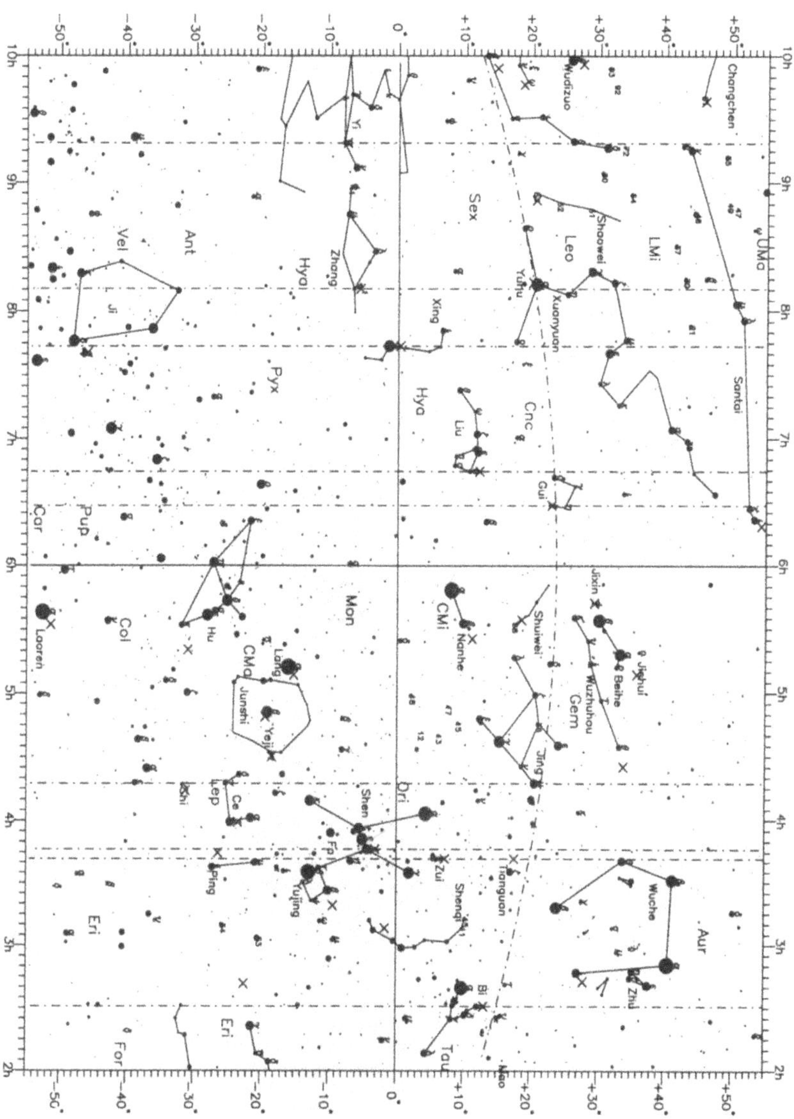

Figure 3.8: The constellations of *Shi Shi*; the Summer quadrant.

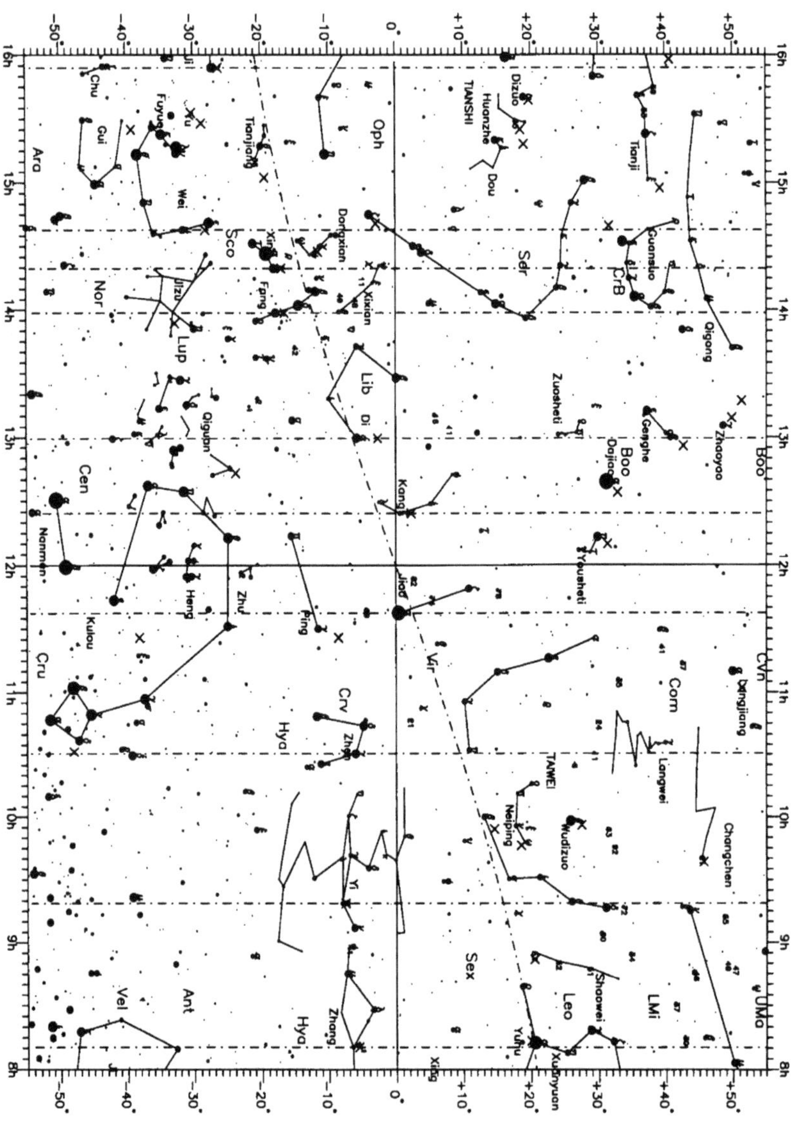

Figure 3.8: The constellations of *Shi Shi*; the Autumn quadrant.

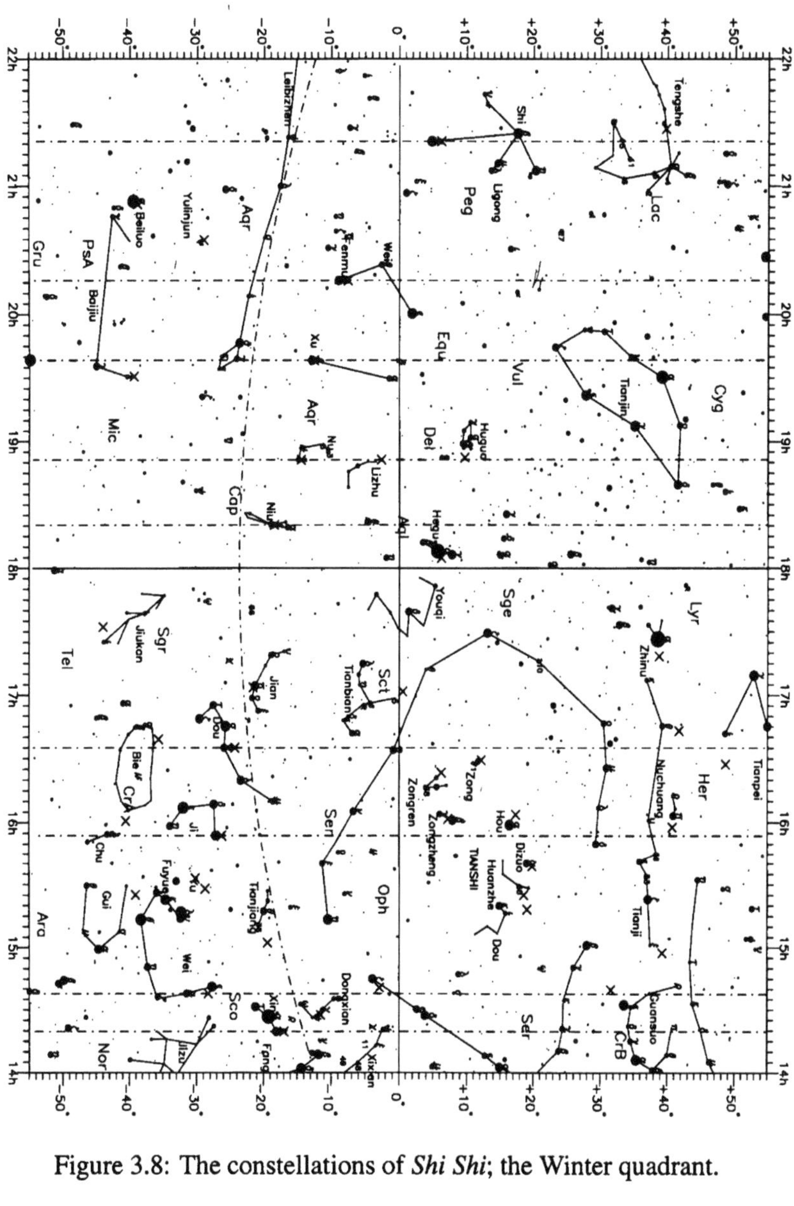

Figure 3.8: The constellations of *Shi Shi*; the Winter quadrant.

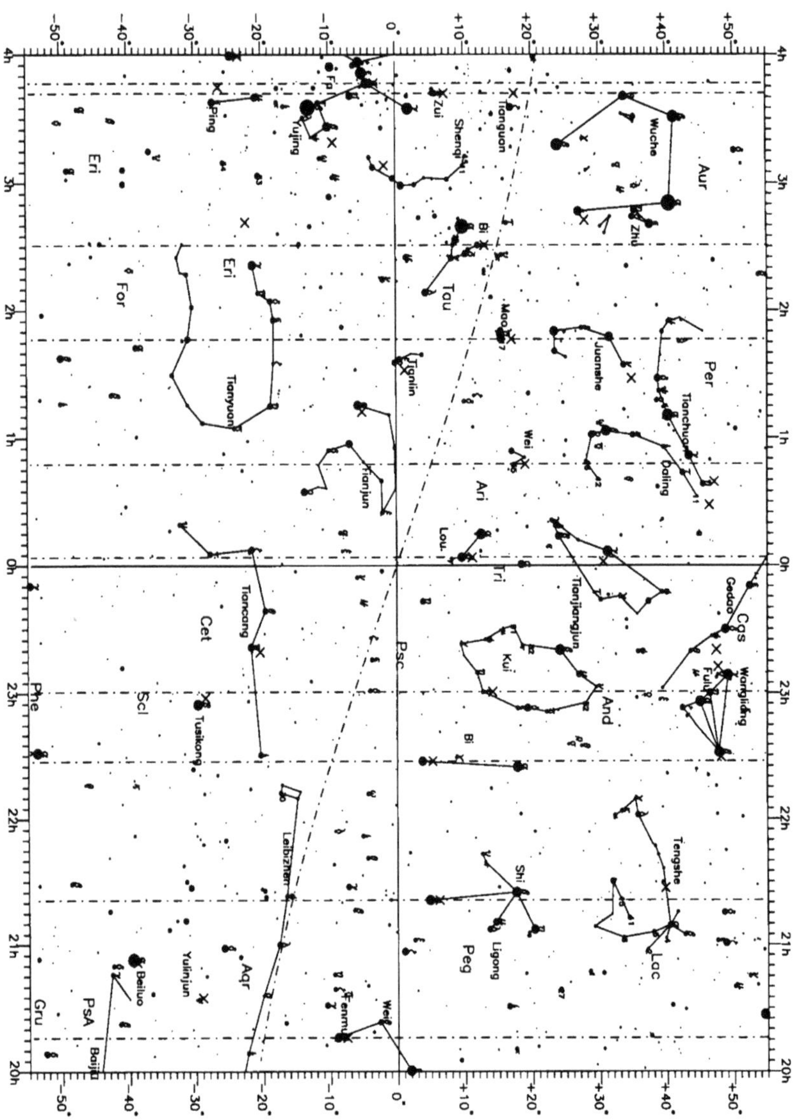

Figure 3.8: The constellations of Shi *shi*; the Spring quadrant.

CHAPTER FOUR

DEVELOPMENT BY GAN SHI AND WUXIAN SHI

4.1 Gan Shi *and* Wuxian Shi

As in the case of *Shi Shi*, we use *Gan Shi* and *Wuxian Shi* to refer to the astronomical schools which made their attributions to Gan De and Wuxian.

4.1.1 Gan De

Gan De was a vague figure in history. It is said that he lived during the Zhanguo period and wrote eight volumes of astrological work. Already during the Han it was not certain where his native place was. Sima Qian said that he was from the State of Qi 齊, nowadays the eastern part of Shandong 山東 Province[1]. Others said he was from the State of Lu 魯, nowadays the area near Mount Tai in Shandong Province[2], or from the State of Chu 楚, present Hubei 湖北 Province[3]. Such confusion could happen during the chaotic period when six dominant states were fighting and competing to obtain supremacy in interstate relations by assuming the title *ba* 霸 (hegemony); a learned person could serve in any of those states. It is even probable that Gan De lived until the end of the Qin dynasty. We found material in the *Han shu* saying that Gan De had given astrological consultation to one of the warlords during the overthrowing of the Qin and the establishment

[1]See the *Tianguan shu*.

[2]In his *Shiji jijie* 史記集解 (Collected Explanations on the *Shi ji*), Pei Ying 裴駰 (c. 400 AD) quoted Xu Guang 徐廣 (c. 400 AD) as saying "The name of Master Gan was De, he was originally from the State of Lu."

[3]See the *Hanshu yiwen zhi* 漢書・藝文志 (Treatise on Bibliography of the *Han shu*).

of the Former Han[4]. Although one might rashly explain this material as a "misplaced string of text," we want to point out that even if Gan De lived until the beginning of the Han it would not be in conflict with the statements that he was a man from the Qi 齊, Lu 魯 or Chu state. It was still customary during the Qin and the early Han for people to identify somebody's native place by means of those, no longer existing, states from the past. On the collapse of the Qin the suppressed ancient aristocrats of the "Six States" rose up against the central government under the banner of re-establishing old states. Geographical and administrative divisions in terms of those ancient states were still in use during the Han. It is quite likely that Gan De lived through the whole Qin dynasty which lasted only 15 years.

The exact contents of the eight volumes of his astrological work are not known to-day. We assume most of it was on planetary astrology. Planetary astrology was very popular during the Zhanguo period; prognostications based on the motion of Jupiter were abundant in books from this period, for example in the *Zuo zhuan* and the *Guo yu*. When Sima Qian made comments in his *Shi ji* on the motion of the planets, he referred to Gan De but not to his constellations. No literature before *Shi ji* mentioned any star names of *Gan Shi*.

Since the Han, however, there appeared many "quotations" from Gan De's work which were indicated by "Gan Shi 甘氏 (Master Gan) said", "Gan Gong 甘公 (Master Gan) said" or "*Gan shi xingjing* 甘氏星經 (Star Manual of Master Gan) said." But the contents of these "quotations" already reflected astronomical and philosophical ideas during the Han. Up to the Sui and the Tang there appeared several books under the name of Gan De. For example, in the chapter on bibliography of the *Sui shu*, a book was listed with the title *Gan shi siqi fa* 甘氏四七法 (Four-times-seven Methods of Master Gan)[5]. Such kinds of books were certainly not written by Gan De himself. The *Tongzhan daxiangli xingjing* which we have mentioned in last chapter is another example.

The case was identical with that of *Shi Shi*. Thus we use the title *Ganshi xingjing* for all astronomical material attributed to Gan De.

---

[4]See the *Han shu*, ch.32, the Biography of Zhang Er 張耳 and Chen Yu 陳餘.

[5]By the title we assume it was probably on the 28 *xiu*.

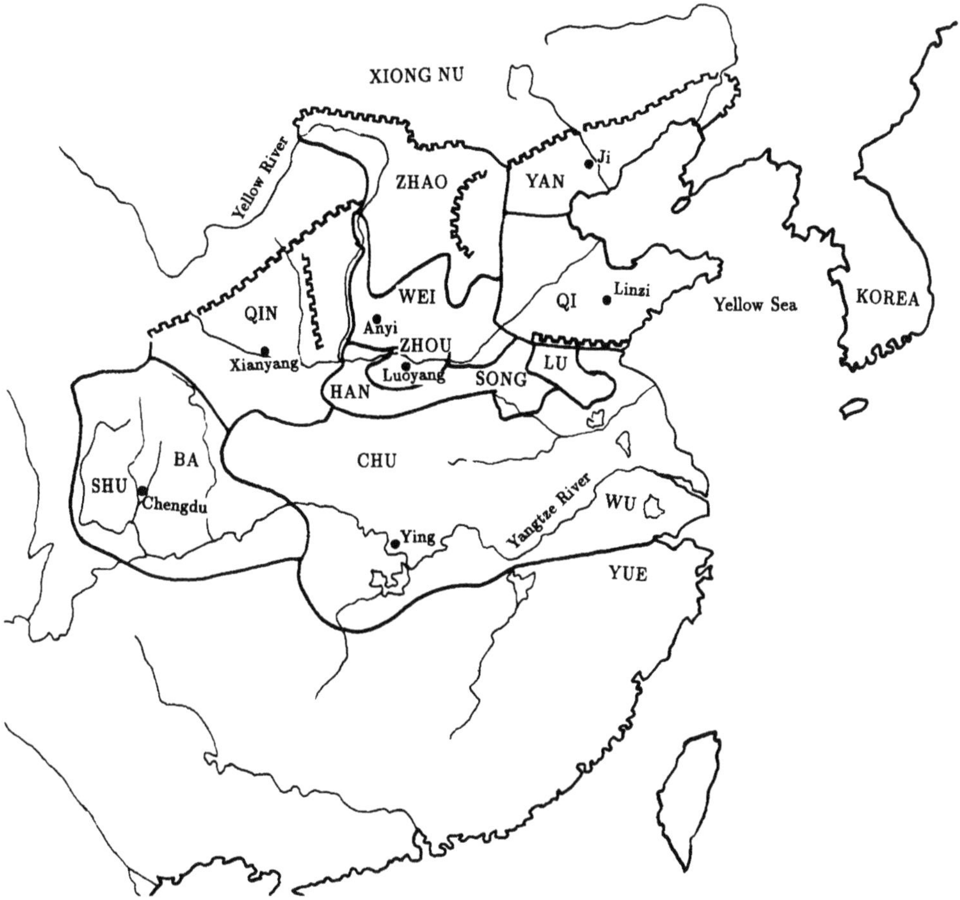

Figure 4.1: The map of China during the Zhanguo 戰國 period (481 - 221 BC).

4.1.2 Wuxian

An astronomer of the Yin dynasty. The *Wuxian Shi* 巫咸氏 made attributions to Wuxian, who was actually a legendary figure. Wuxian was mentioned as an astronomer by Sima Qian in his *Shi ji.* He said in the postscript of the *Tianguan shu:*

> Counting the astronomers who passed on the knowledge on the sky we have: Chong 重 and Li 黎 before Emperor Gao Xin 高辛; Xi 羲 and He 和 for Emperor Tang 唐 and Yu 虞 ; Gun Wu 昆吾 for the Xia 夏 dynasty; Wuxian for the Yin-Shang 殷

> 商 dynasty; Shi Yi 史佚 and Zhang Hong 萇弘 for the royal family of the Zhou 周; Zi Wei 子韋 for the state of Song 宋; Bi Zhao 裨竈 for the state of Zheng 鄭; Master Gan (De) for the state of Qi 齊; Tang Mei 唐眛 for the state of Chu 楚; Yin Gao 尹臯for the state of Zhao 趙; (and) Master Shi Shen for the state of Wei 魏.

All the mentioned astronomers before the Zhou seem to have been mythological or legendary persons.

Wuxian was also said to be a "virtuous minister" at the time of Emperor Da Wu 大戊 of the Yin dynasty about 1200 BC[6]. Moreover he was said to be the inventor of divination with yarrow sticks. It is quite probable that once there existed a person Wuxian in history, but in literature from the Qin and Han onwards, Wuxian became a fictitious personage representing ancient shamans[7].

It was not until the Later Han that some astrologers began to write in the name of Wuxian. Then the *Wuxian Shi* astronomical school came into existence[8]

Chen Zhuo and Wuxian Shi Because it was Chen Zhuo who finally fixed the constellations of the three schools, and because no quotations about *Wuxian Shi* are available as there are in the case of *Shi Shi* or *Gan Shi* in the Han literature, it seems likely that Chen Zhuo himself wrote in the name of *Wuxian Shi*.

Chen Zhuo lived during the Sanguo period and the beginning of the Jin 晉 dynasty, from 220 to 300 AD. He took the position of *taishi ling* 太史令 in the Kingdom of Wu before 280 AD and afterwards at the royal Jin court. He wrote nine books on astronomy and one of them was on the constellations of the three schools of *Gan Shi*, *Shi Shi* and *Wuxian Shi* (according to the *Sui shu*). Actually Chen Zhuo's ancestral home was in the Kingdom of Wu (Capital: Jian Ye 建業, present day Suzhou), the same

---

[6]See the *Shi ji*, ch. 3, *yin benji* 殷本紀 (Biographies of the Yin Kings).

[7]See Jiang Xiaoyuan 江曉原, *Tianxue zhenyuan* 天學眞原 (Prime Themes of Ancient Learning of Heavens; Liaoning, 1991), pp.83-89.

[8]Qian Baocong 錢寶琮, "Ganshi xingjing yuanliu kao" 甘石星經源流考 (The origin of the *Gan Shi xingjing*), *Guoli zhejiang daxue jikan* 國立浙江大學季刊 (Quarterly Publication of the State University of Zhejiang), 6 (1937).

as that of the legendary Wuxian[9]. This provided additional support for the assertion of Chen Zhuo's relation with *Wuxian Shi.*

Recently, Pan Nai [1989] discovered in the *Tianwen yaolu* from the Tang ten values of the distances to the pole (QJD) of the *ju xing* of ten *xiu*. These ten values were headed with "Wuxian of the Yin said." Pan Nai's analysis of these data revealed a date of about 255 AD[10]. This is probably another proof of Chen Zhuo's dedication to *Wuxian Shi.* So we have some reason to believe that Chen Zhuo was the leader of *Wuxian Shi* of his time.

The constellations of *Wuxian Shi*, however, might have already existed before the time of Chen Zhuo. This question needs to be investigated.

4.2 Demarcation of the constellations of the three schools

Although Chen Zhuo's allocation of the constellations to the three schools has been accepted by astronomers of later times, it would be erroneous to believe that each of these groups of constellations were exclusively the constellations of each of the three schools before Chen Zhuo, or during the Han. It seems incredible to assume that *Wuxian Shi,* for example, studied only 44 small constellations in the sky without looking at others. Most probably all three schools investigated the total sky and some constellations were studied by all three. That would mean that the constellations of the three overlapped. What we believe to be the groups of constellations specifically attributed to one of the three schools was only Chen Zhuo's allocation. "He *conflated* the constellations of the three schools and got 283 constellations including 1464 stars."[11]. The actual situation during the Han needs to be studied. We will try to find out how these three groups overlapped.

No direct reference exists to tell us which constellation belonged to which school during the Han. But there is one kind of information which we can use for our purpose, namely, the mutual astrological commentaries on the constellations by the three *Shi* 氏. These commentaries have been

[9]The *Suzhou fuzhi* 蘇州府志 (Local History of Suzhou; compiled in the later Qing dynasty) says: "the tomb of Wuxian is located three *li* 里 (one *li* equals half a kilometer) northeast to the Ping gate of Suzhou; he was the virtuous minister from the time of Da Wu 大戊 of the Shang 商 (Yin) dynasty." Quoted from Pan Nan [1989]; p.116.

[10]Pan Nai, *Zhongguo hengxing guanche shi*, 1989, pp. 116-117.

[11]See *Jinshu tianwen zhi.*

quoted in astrological books after the Han. At the beginning of the Tang dynasty there were many books of this kind; the *Kaiyuan zhanjing* was the most complete one. In the *Kaiyuan zhanjing* one can see that on some constellations of one school, the other two made astrological commentaries as well. For example, on the star *Dajiao* 大角 of *Shi Shi,* both *Gan Shi* and *Wuxian Shi* gave their comments; it is evident then that *Dajiao* was also studied by *Gan Shi* and *Wuxian Shi* and it did not exclusively belong to *Shi Shi.*

By collecting statistics of the comments by the three schools we get an estimate of how many constellations each school studied during the Han. The first statistics of this kind were collected by Liu Jinyi 劉金沂 and Wang Jianmin 王健民 based on the *Kaiyuan zhanjing*[12]. We checked their results by looking through two more books from the Tang time: *Tiandi ruixiang zhi* and *Tianwen yaolu.*

Our statistics gives the following results:

— of 92 constellations of *Shi Shi,* 61 have also been commented on by *Gan Shi;* and 56 by *Wuxian Shi;*

— of 118 constellations of the *Gan Shi,* 11 have also been commented on by the *Shi Shi;* and 19 by the *Wuxian Shi*;

— of 44 constellations of *Wuxian Shi,* 4 have also been commented on by *Shi Shi;* and none by *Gan Shi.*

The 28 *xiu* were the basis of the Chinese sky; they certainly were commented on by all schools[13].

We illustrate this result with Figure 4.2. The three thick-lined squares on the left side of this figure represent the three schools: "S" for *Shi Shi,* "G" for *Gan Shi* and "W" for *Wuxian Shi.* The numbers inside indicate the number of constellations ascribed to each school according to Chen Zhuo.

But each of these schools also referred in their commentaries to constellations of the other two schools. The number of constellations has been

[12]Liu Jinyi 劉金沂 and Wang Jianmin 王健民, "Chen Zhuo he Gan Shi Wu sanjia xingguan" 陳卓和甘石巫三家星官 (Chen Zhuo and Constellations of Gan, Shi, and Wu), *Kejishi wenji* 科技史文集 (Collected Papers on History of Science), 6 (1979), pp.32-44.

[13]There were some deviations in the names of the *xiu* and even about the *xiu* constellations. In the *Shi ji* one can find two slightly different systems of 28 *xiu.* One system contained the ordinary 28 *xiu* which were adopted by *Shi Shi;* the other system had the name *Zhu* 注 for *xiu Liu* 柳, and used the constellations *Lang* 狼 and *Hu* 弧 for *xiu Jing* 井 and *xiu Gui* 鬼. The latter was apparently adopted by *Gan Shi.* See Qian Baocong [1947].

S: 92	G: 11	W: 4
The 28 xiu		

G: 118	S: 61	W: 0
The 28 xiu		

W: 44	S: 56	G: 19
The 28 xiu		

Figure 4.2: The demarcation of the constellations of the three schools. "S" stands for *Shi Shi;* "G" for *Gan Shi;* "W" for *Wuxian Shi.* See text.

visualized in the two thin-lined squares to the right of each school. So we see, for instance, that *Shi Shi* with its 92 constellations also considered and discussed 11 constellations of *Gan Shi* and 4 of *Wuxian Shi.*

Still the exact situation can not be known from such statistics. Firstly the completeness of the quotations in the Tang books was not guaranteed; secondly some reputed quotations might originate from after the Han. So here we can only provide an estimate of the situation.

It is even more interesting to look at the number of constellations of one *Shi* which were commented by the other two. Since comments always came after the construction of the constellations, our results also indicate the relative ages of the constellations of the three schools. Most *Shi Shi* constellations have been commented on by *Gan Shi* and *Wuxian Shi*, but quite contrarily, only a few constellations of *Gan Shi* and *Wuxian Shi* have been commented on by *Shi Shi.* Thus the system of *Shi Shi* must be the oldest. By the same reasoning the system of *Gan Shi* was older than that of *Wuxian Shi.* So the system of *Shi Shi* apparently served as the framework for both *Gan Shi* and *Wuxian Shi.* Both schools further developed the Chinese sky within the original framework of *Shi Shi.* Our further study confirms this point.

4.3 Constellations of Gan Shi

According to Chen Zhuo, 118 constellations belonged to *Gan Shi.* Contrary to the case of *Shi Shi*'s constellations, no coordinates for the leading stars were found in the *GSXJ.* Actually no leading stars were mentioned at all. The descriptions in the *GSXJ* indicated the positions of the constellations of *Gan Shi* directly or indirectly relative to the constellations of *Shi Shi.* According to these descriptions and with the help of star maps from the Tang and the Song one can get a general perspective of the constellations which *Gan Shi* constructed.

4.3.1 Ideological constructions

Our strongest impression is that the constellations were not constructed with a view of positional astronomy. They were ideological constructions based on tradition and imagination. Some constellations were composed of faint stars, some of them so faint as to be almost invisible. Apparently the exact position of these constellations was not the interest of Gan De 's school. Most important for them was the indication of certain entities or concepts in the sky.

Before going into details in the following chapters we would like to take the constellations of *Gan Shi* inside the *Ziwei Yuan* 紫微垣 (Purple Court) in the north polar area as an example to demonstrate ideological constructions in the sky.

The *Ziwei Yuan* was supposed to be the celestial Imperial Court. *Gan Shi* furnished it with the constellations:

> *Tianhuang dadi* 天皇大帝 – *Sifu* 四輔 – *Yinde* 陰德 – *Liujia* 六甲 – *Shangshu* 尚書 – *Nüshi* 女史 – *Tianzhu* 天柱 – *Zhushi* 柱史 – *Huagai* 華蓋 – *Tianchuang* 天床 – *Wudi neizuo* 五帝内座[14].

[14]See Appendix II for the meaning of these star names.

In Appendix II we will try to explain all these star names, and find that they all correspond with the Imperial Court. Here let us only take *Tianhuang dadi* as an example to illustrate how the development of ideology influenced the naming of stars. However one should not expect a critical historical or sinological study of the characters used in this denomination of stars here.

During the Zhou, the belief in the Mandate of Heaven (*tian ming* 天命) was highly regarded and propagated widely. The Zhou rulers claimed to have received the Mandate of Heaven to succeed the Yin rulers. The God issuing the Mandate was called *Tian* 天 (Heaven) or *Di* 帝 (God or Emperor), with the latter being very frequently referred to as *Shangdi* 上帝 (Upper God). *Tian* was the ultimate God who presided over all the other gods and all human affairs and natural phenomena.

During the Yin, such ideas about the Mandate of Heaven had certainly existed. But by the divine inscriptions on oracle bones, it seems that the character *Di* had been used to indicate this ultimate God, and not *Tian*. Whatever these people were doing they consulted *Di* by scapulimancy.

During the early Zhou this *Tian* or *Di* was also referred to as *Tian Di* 天帝. The old Chinese character *Tian* was composed in such a way as indicating the sky above a figure of man[15]. Apparently already then the sky was identified with the awesome, high God of Heaven *Tian Di* 天帝, as one can also find among other Asiatic peoples[16].

Then came the hierarchical question of which Being in the sky could take care of terrestrial life as a representative of Heaven. For most civilizations that Being is the Sun, or rather the light or warmth of the Sun which stimulates all life. According to the *Yaodian* 堯典 (Canon of Yao), the brethern Xi 羲 and He 和 were sent by Emperor Yao to the East and to the West to welcome the Sun on rising and to take leave of it on setting. This was actually a reflection of the worship of the solar god.

The Chinese character for the Emperor *huang* 皇 can phonetically mean Yellow (*huang* 黃), the colour of the Sun. It can also mean the Brilliance

---

[15]This way of composing the Chinese characters is called *zhi shi* 指事 the self-explanatory method.

[16]For the Babylonian pantheon of about 1800 BC see A. Heidel [1942], pp. 16-44. The pantheon of Nippon is represented in the first chapter of *Kojiki;* see D. Philipi [1968]. The Mongolian pantheon is described in their *Tobchi'an*; see W. Heissig [1980] and F.W. Cleaves [1982].

(*huang* 煌) of the Sun. Apparently the Emperor (*huang* 皇) was directly associated with the Sun; he was the Sun of human society. The combination of *Huang* 皇 with *Tian* 天 or *Di* 帝 produced terms like *Tian Huang* 天皇 (Heavenly Emperor), *Huang Tian* 皇天 (Imperial Heaven) or *Huang Di* 皇帝 (Imperial Emperor)[17] which became a specific title for the terrestial emperor after Qin Shi Huangdi. In all these terms we observe the dominant position of the solar god in the hierarchical structure of Heaven.

At the end of the Zhou a tendency developed to consider the celestial pole as the highest representative of *Tian.* Circumpolar stars like *Beidou* (UMa) moving around the pole were one of the mysterious phenomena to ancient people. Actually the movement of *Beidou* around the pole had been used since ancient times to indicate the seasons. This fixed abstract point, the celestial pole, was highly regarded by ancient Chinese philosophers. Confucius claimed that

> He who is a governor because of his Virtue may be compared with the North Pole Star, sticking to its place and all stars turning about it.[18]

After Confucius the academic tendency was to strive for the ultimate Unity and Harmony in the Universe. According to the Confucian tradition, the terrestrial Ultimate Sovereign should maintain Order and Harmony in his empire like the celestial Pole in its cosmic surroundings.

According to the philosophy of Taoist school, which sought an abstruse and ultimate *Dao* 道(The Way) of the universe, the *Dao* produces One (*Yi* 一) which in turn produces all things in the universe. Probably inspired by this philosophy, the *Yin-yang* philosophers combined the *yin* and *yang* forces to form an initial Supreme One, *Tianyi* 天一 or *Taiyi* 太一. The interaction of *yin* and *yang* on the cosmic fluidum *qi* 氣 led to the creation of the Five Elements (*wu xing* 五行) and all other things.

The celestial pole provided a common answer for all the inquiries of philosophical schools. It was synonymous with the Celestial Emperor for Confucians, with the Supreme One (*Taiyi* 太一) for the *Yin-yang* philoso-

---

[17]For a discussion about the meanings of these homonyms *Huang* (皇,黄,煌), see W. Münke's study on *Huangdi* 黄帝 (Yellow Emperor); *Die Klassische Chinesische Mythologie,* Ernst Klett Verlag. Stuttgart, 1976, PP.161-174.

[18]*Lun yu* 論語 (Confucian Analects); translation taken from Legge [1861].

phers and with the Source of Ultimate *Dao* 道 for the Taoists. During the Han, the Taoists produced the ultimate deity called *Tianhuang dadi* 天皇大帝 which was simply a combination of *Tian* 天, *Huang* 皇, *Da* 大 and *Di* 帝, all representing the God of Heaven. He was so high, so abstruse and so mysterious that he could only be represented in the sky as a hardly visible star inside *Ziwei* 紫微 (near α UMi) — God acts but is not visible.

Obviously the constellations named in *Ziwei* by *Gan Shi* originated from the Han schools and academies. They represented officers, some material objects and even some abstract esoteric concepts according to the Taoist philosophy.

The constellation *Shangshu* 尚書 probably directly indicates the time when constellations in *Ziwei* were created by *Gan Shi*. The celestial construction of four secretarial offices plus a supervisor (in total five stars) probably corresponds with officialdom during the reign of Emperor Cheng 成 (32 - 7 BC), who introduced this innovation inside the Royal Palace in Chang'an[19]. See Appendix II, no.9.

4.3.2 Historical investigation of *Gan Shi*'s constellations

Because no numerical data are available for the *Gan Shi* constellations, textual research is necessary. In textual research, one must inevitably sort through details of historical evidence in order to arrive at the best possible conclusions. We present all our textual research and explanations concerning *Gan Shi*'s constellations in Appendix II and do the same for *Wuxian Shi* in Appendix III. There one can find:

— *the names of stars and constellations* in *pinyin* spelling and in Chinese characters together with a brief English translation.

— *the number of stars* given in the *KYZJ*, a brief description of *their positions* according to the *GSXJ* and our identifications.

— *some explanations and comments on the meaning of star names*, based on textual research or in some cases based on historical evidence and cultural context.

Our identifications were of course based on historical maps, star catalogues and former studies as far as available; but one has to accept that

[19] See the *Tong dian* 通典; this book is a reservoir of source material on political and social history which was compiled by Du You 杜佑 about 812 AD.

our knowledge about the constellations of *Gan Shi* is more speculative than about those of *Shi Shi*. The same holds for the constellations of *Wuxian Shi*. Our preliminary results might have some value, however, for cultural and astro-mythological studies.

4.4 *Constellations of* Wuxian Shi

According to Chen Zhuo there were 44 constellations in the sky which belonged to *Wuxian Shi*. In comparison with those of the other two schools they were less important, however. The 44 constellations consisted of only about 140 single stars, mostly rather faint. Like the constellations of *Gan Shi*, those of *Wuxian Shi* filled some vacancies left over by *Shi Shi*. Moreover, the number of constellations of *Wuxian Shi* is so limited that one can hardly find any system in it, unlike in the case of *Gan Shi*. We notice some special points.

— On the market-square within *Tianshi Yuan* 天市垣 we see *Liesi* 列肆 (small shops), *Chesi* 車肆 (wagon shop), *Bodu* 帛度 (textile ruler; indicating cloth vendors) and *Tusi* 屠肆 (butcher's shop), as aspects of the Market. They reflect the traditional market folklore.

— In the sector of *Qiguan* 騎官 (cavalry officers) of *Shi Shi*, we notice the addition of *Dunwan* 頓頑 (military officer to maintain discipline) and *Congguan* 從官 (military doctor[20]).

— Other constellations belong typically to the category of filling functional vacancies within *Shi Shi*'s framework like

Dali 大理, the Great Judge belonging to the Court of *Ziwei* 紫微;

Huben 虎賁, brave soldiers guarding the Court of *Taiwei* 太微;

Tianfu 天桴, a celestial drumstick for the big drum *Hegu* 河鼓 at the River.

For the constellations of *Wuxian Shi*, we followed the same procedure as for *Gan Shi*. Textual research and explanations are given in Appendix III. The constellations were reconstructed and presented together with those of *Shi Shi* and *Gan Shi* on one final map for the epoch of 100 BC.

[20] *Congguan* 從官 literally means Assistant Officer or Page. Our explanation is based on the celestial context and astrological descriptions in the *Kaiyuan zhanjing* (chapter 70).

4.4.1 When were *Wuxian Shi*'s constellations named?

We did some research on this problem and discovered that it happened before the astrologer Xi Meng 郗萌 was active.

In the *Kaiyuan zhanjing*, quotations from various authors and books from the Han time have been compiled into paragraphs with names of constellations as their titles. Besides "quotations" from the three schools[21], *Kaiyuan zhanjing* also quoted some famous historical persons. One of the most frequently quoted authors was Xi Meng.

There are two more astrological collections dating from the Tang, compiled in the same way. They are

— *Tianwen yaolu* 天文要錄 (Essentials of Astrology) compiled by Li Feng 李鳳 in 664 AD.

— *Tiandi ruixiang zhi* 天地瑞祥志 (Records of Omina from Heaven and Earth) by Sa Shouzhen 薩守眞 in 666 AD.

Both books are no longer complete. The remaining parts are nowadays in Japan. In the Beijing Library copies of these preserved parts are available.

We did some statistical research on these three sources and discovered that 18 of *Wuxian Shi*'s constellations, out of a total of 33[22], have been commented on by Xi Meng. Considering that these were only quotations and that two books from which we derived the statistics were not complete, we are convinced that the constellations of *Wuxian Shi* existed before Xi Meng.

The question remains, who was Xi Meng? Up to now we know of him from the *Jin shu tianwen zhi*. Li Chunfeng called him a "Gentleman Librarian" (*bi shu lang* 秘書郎). He lived during the Later Han and taught about the *xuan ye* 宣夜 ("Darkness of the Void") theory, one of the three most popular cosmological theories.

In the *KYZJ* we meet frequent quotations about Xi Meng's teachings on various astronomical subjects. Nevertheless, all these references and quotations do not tell us exactly when he lived.

[21]Each paragraph, marked as mentioned above by the names of a constellation, began with the name of the school to which this constellations belonged.

[22]The twelve constellations of twelve States of *Wuxian Shi* were treated as one collective constellation called *Shi'erguo* 十二國 (Twelve States) in the *Tianwen yaolu* and *Tiandi ruixiang zhi*. Hence the number 33 and not 44.

This situation changed substantially by our discovery of a decisive quotation in an ancient book from the early fifth century. This book is *Wen xuan* 文選 (General Anthology of Prose and Verse) edited by Xiao Tong 蕭統, a prince from the Liang 梁 dynasty (502 - 557). It contains pieces of literature from various authors from most ancient times until the time of compilation of the *Wen xuan*. Among this collection is a paragraph by Ban Gu 班固 (32 - 92), the author of the *Han shu*. It reads as follows:

> In the 17th year of Yongping 永平 (74 AD) I, the subordinate of Your Majesty, and Jia Kui 賈逵, Fu Yi 傅毅, Du Qu 杜矩, Zhan Long 展隆, Xi Meng 郗萌, and several others met at Yunlongmen 玉龍門. *Xiao huangmen* 小黄門 (Junior Secretary) Zhao Xuan 趙宣 presided the meeting. ...

Among these persons, Jia Kui 賈逵 is known to historians. He was a famous philosopher and astronomer of the Later Han who lived between 30 and 101 AD. Now we know Xi Meng was a contemporary of Ban Gu and Jia Kui and was active in the year 74 AD.[23]

From the above historical study we conclude that most constellations of *Wuxian Shi* were named later than those of *Shi Shi* but earlier than 74 AD. They were the product of the same academic atmosphere in which the constellations of *Gan Shi* were created.

4.5 *Reconstruction of the constellations of the two schools*

4.5.1 Useful material for our reconstruction

In most cases the identification of Gan De's constellations is possible according to the descriptions in the *GSXJ* text which provided fundamental information. For the precise shape of some constellations, we have to refer to existing historical star maps and former studies by others. So in the reconstruction of *Gan Shi*'s sky we have condensed our historical knowledge.

After the reconstruction of *Shi Shi*'s constellations, the reconstruction of *Gan Shi*'s constellations is a process of filling the open places within the

[23] J. Needham estimated that Xi Meng was a younger contemporary of Zhang Heng 張衡 (78 - 139). See *Science and Civilization in China*, 1959, vol.3, section 20, p.219.

framework made by Shi Shen's school. We used the following material for this effort.

Texts

1) *Jin shu tianwen zhi* 晉書・天文志 and *Sui shu tianwen zhi* 隋書・天文志: these astronomical chapters of the history of the Jin and Sui dynasties were all written by Li Chunfeng 李淳風.

2) *Kaiyuan zhanjing* 開元占經, the chapters 69, 70, and 71, all devoted to the constellations of *Gan Shi* and *Wuxian Shi.*

3) *Tongzhan daxiangli xingjing* 通占大象曆星經, the so called *Ganshi xingjing* 甘石星經.

4) *Bu tian ge* 步天歌 by Wang Ximing 王希明 of the early Tang.

5) Material concerning the constellations of the three schools in the Dunhuang 敦煌 Manuscripts and some other books from the Tang: *Tiandi ruixiang zhi* 天地瑞祥志 and *Tianwen yaolu* 天文要錄.

Among the above five kinds of source material, the first two are the most important. The latter three were used for comparison. Essentially there were not many deviations from the first two. All these books were from the Sui and the Tang or a little bit earlier (500 - 700). But all of them were most probably based on Chen Zhuo's work.

Star maps and catalogues

a). Dunhuang star maps.

This is the earliest existing set of star maps of the whole sky in China[24]. They were discovered in the Buddha Arches of Dunhuang, Gansu. Although they were by no means a precise scientific star atlas, they presented the Chinese sky in a fairly recognizable way. They contain one separate part for the northern circumpolar stars and twelve parts for the twelve months.

[24]Two star maps earlier than the Dunhuang map have been discovered: one is the Jiaotong star map on the ceiling of the Han tomb; the other is the Loyang star map also painted on a tomb ceiling, dating from 526 AD. But as far as positions of stars are concerned, both star maps are so primitive that they have almost no value for the identification of ancient constellations. See R. Stephenson; "Stargazers of the Orient," *New Scientist*, 1(1993), pp. 32-34.

Almost all constellations can be found on these maps and the constellations of different schools were marked with different colours: yellow for *Shi Shi* and *Wuxian Shi* and black for *Gan Shi*. So this set of maps is a valuable reference for the study of the constellations of the three schools.

b). Song maps and catalogues

1). The *Xin yixiang fayao* 新儀象法要 star maps.

This star map of Su Song 蘇頌 (1020 - 1101) is the earliest known star map which was presented in a scientific way. The technique of measurements had developed to a high level during the Song. At least four well organized series of observations were carried out. Su Song's map was based on observations during the Yuanfeng reign (1082). The relative positions of the constellations fit quite well to reality. Constellations belonging to different schools were also distinguished by black dots and by open circles; open circles for *Shi Shi* and *Wuxian Shi*, black dots for *Gan Shi*. These features of Su Song's map proved to be a great help for our study of the constellations of the three schools.

2). The *Suzhou tianwen tu* 蘇州天文圖.

This is the best known star map from Chinese history. It was engraved on a tablet of stone by Wang Zhiyuan 王致遠 in the year of 1247 according to the design by Huang Shang 黄裳 (1147 - 1195). Investigations have shown that the positions of the stars on this map are rather accurate[25]. This astronomical chart was based on observations in the 11th century, although the indicated position of the equinoxes corresponds with the Han-time (100 BC). *Suzhou tianwen tu* is an important landmark from where we can trace back to the Chinese sky before the Song.

3). The Huangyou 皇祐 Star Catalogue

This catalogue was based on observations during the Huangyou period of the Song dynasty. For the first time the coordinates of the leading stars of the constellations of *Gan Shi* and *Wuxian Shi* were given. Thus this catalogue gives valuable indications for the reconstruction of the constellations of these two schools. Pan Nai [1989] made a thorough study of this catalogue. He even made a reconstruction of the Song sky based on the

---

[25]Du Shengyun 杜升雲 [1982] measured the coordinates of 266 stars from the *Tianwen tu* and compared them with the calculated coordinates for that time and revealed the square root variance was 1.5 degrees.

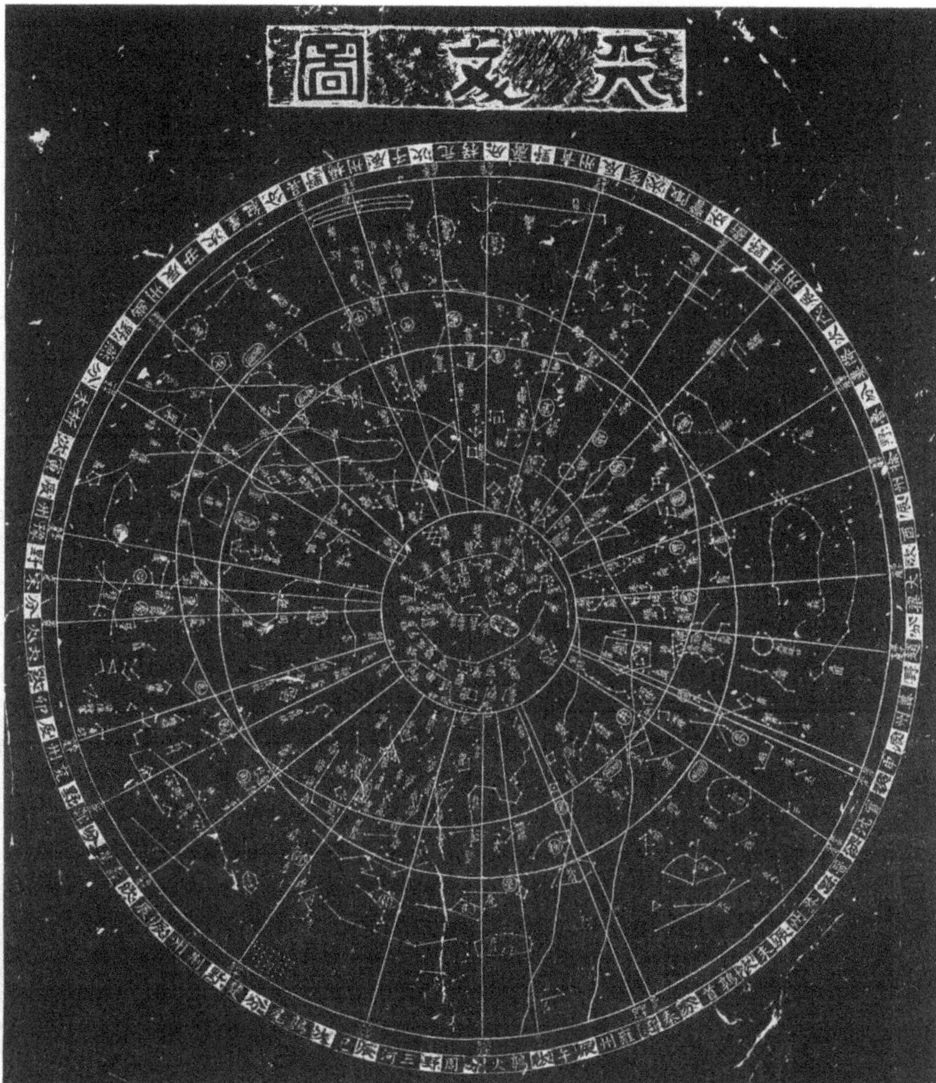

Figure 4.3: *Suzhou tianwen tu* 蘇州天文圖, the Astronomical Planisphere in Suzhou (1247 AD).

Huangyou catalogue. The Huangyou data are also used by us, see Appendix II.

c). The *Tianwen huichao* 天文彙鈔 Star Catalogue from the Ming.

A star catalogue together with some illustrative star maps was recently discovered in a Ming hand-copy of a collection of books. Chen Ying 陳鷹 [1986] and Pan Nai 潘鼐 [1989] both made independent studies and both arrived at the same conclusion that this star catalogue was observed by Guo Shoujing at the beginning of the Yuan. But our examination by means of Fourier analysis of the data in the catalogue proved that this catalogue was observed in the beginning of the Ming, probably near the capital Nanjing. Coordinates of 740 stars of 200 constellations are available in this catalogue and the accuracy of the data is about half a degree. This star catalogue provides another important landmark on the way of tracing back towards the ancient Chinese sky.

d). The *Chongzhen lishu* 崇禎曆書 star catalogue and atlas.

This catalogue and star atlas were part of the *Chongzhen lishu* (Calendrical Encyclopedia of the Chongzhen Period) by Xu Guangqi 徐光啓 and Adam Schall von Bell (Tang Ruowang 湯若望) at the turn of the Ming and Qing dynasties. This work introduced western astronomy in China. All the ancient Chinese constellations were observed using western methods. European star catalogues and star atlasses were absorbed in this tremendous work[26]. The Chinese constellations were put in an accurately measured celestial background. Their most valuable star atlas is the *Chidao nanbei liangzong xingtu* (General Star Atlas of the Southern and Northern Hemispheres Separated by the Equator)[27]. Their way of presenting the Chinese sky was more scientific than in earlier times. Faint and hardly recognizable stars, which had been traditionally included during many centuries in the Chinese constellations, were left open in their maps and catalogues. For the constellation *Huagai* 華蓋 of 16 stars by *Gan Shi*, for example, they only indicated 3 clearly visible stars. The star atlas by Xu Guangqi and Schall

[26]See D'Elia, "The double stellar hemisphere of Johann Schall von Bell, S.J.", *Monumenta Serica*, 1959; and Keizo Hashimoto, *Hsü Kuang-Ch'i and Astronomical Reform*, Kansai University Press, 1988, p.162.

[27]We want to thank professor Ulrich Libbrecht and Dr. Grete Moortgat from Leuven for helping us with a slide copy of this atlas preserved in the Vatican Library in Rome. This made it possible for us to examine the detailed configurations of constellations in the atlas.

von Bell was the result of the first cooperation between western and Chinese astronomers. Therefore it deserves historical importance in our search for the Ancient Chinese sky. It is one of the starting points of our research from which we try to reach the Han times.

4.5.2 Conventions in our reconstruction

We do not expect the presentation of the constellations on our maps to be considered precise for each individual star. As a matter of fact, the coordinates of most of the 1464 individual stars of 283 constellations did not exist before 1673 when the Belgian Jesuit Ferdinand Verbiest (Nan Huairen 南懷仁) compiled the *Lingtai yixiang zhi* 靈臺儀象志 (Notes on the Astronomical Instruments in the Imperial Observatory) in which the coordinates of about 1800 stars were given for the first time. Almost all the 1464 individual stars of the classical Chinese sky were included. But the *Lingtai yixiang zhi* should be considered only as a Qing version of the sky. Indeed, for some constellations of *Gan Shi* (and *Wuxian Shi*) it is impossible to identify every individual star. From the very beginning, these constellations were vague projections in the sky of rather abstract notions. Their real footholds were usually one or two visible stars in the area of their projected fantasies.

We maintain the following conventions in our reconstruction of *Gan Shi*'s and *Wuxian Shi*'s sky.

1) The location of a reconstruction must be in accordance with the description of its position given in the text of the *Jin shu* and the *KYZJ*. Our star map for the Han-epoch based on the *SSXJ* exact coordinates has been indispensable for this work. The star map for the Han epoch is so helpful because relative orientations between constellations eventually changed appreciably due to two thousand years of precession, especially at high declinations. The *GSXJ* text only gave the relative orientations of the constellations.

2) When the description in the *GSXJ* text was not enough to make our choice we used other sources. Priority was given according to the antiquity of the source. *Suzhou tianwen tu* and the Star Catalogue in the *Tianwen huichao* are the most reliable references.

3) For the shape of the constellations we adopted Yi Shitong's design if we had no earlier and convincing reference. In general, Chinese constella-

tions exist of sketches of straight lines connecting the constituent stars.

Our final reconstruction of the *Gan Shi* and *Wuxian Shi* constellations was presented in a second set of star maps at the end of this book, together with the constellations of *Shi Shi*. So this second set of star maps is the complete reconstruction of the Chinese sky during the Han.

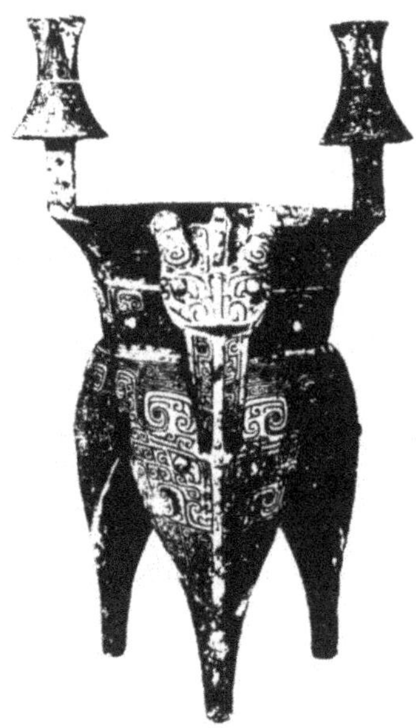

Bronze clepsydra (water-clock) from the Former Han, excavated in Ikchao, Inner Mongolia, China. The water-clock was invented during the early Zhou, before 770 BC. It was essential for astronomical observations for calendar-making. Photo taken from the *Zhongguo gudai tianwen wenwu tuji* (Wenwu Press, Beijing; 1980).

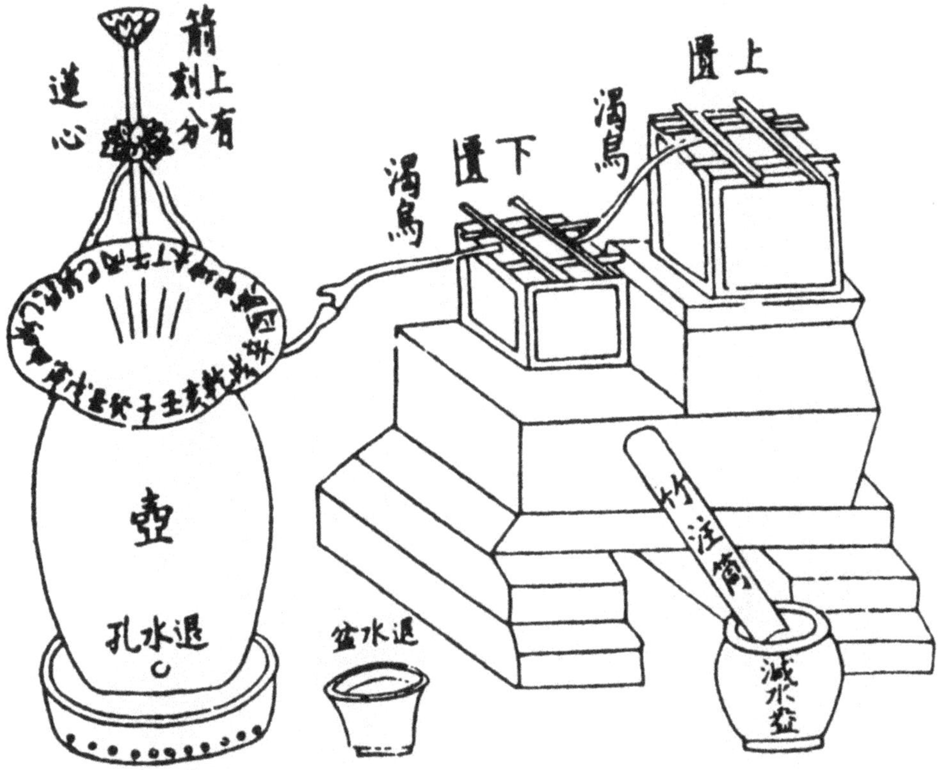

A polyvascular inflow clepsydra made by Yan Su (c. 1030 AD) of the Song dynasty. The upper vessel is the reservoir. The second vessel maintains a fixed water level by adoption of an overflow principle, so it provides a steady water flow to the receiver. In the receiver a calibrated floating "arrow" indicates the water level, telling time. A water-clock was essential for accurate astronomical measurements.

Picture extracted from the *Liujing tu* by Yang Jia (c. 1155 AD) of the Song.

CHAPTER FIVE

PHILOSOPHY OF THE CHINESE SKY

5.1 A reflection of human society

So far we have acquired a complete representation of the Chinese sky by investigating the constellations of the three astronomical schools during the Han. The sky was composed of 283 constellations, 121 of them (including the 28 *xiu*) were described by Shi Shen's school. Some of them obviously originated from ancient times long before the Han, but were not systematically measured and registered until the Former Han. The constellations of Gan De's and Wuxian's schools were later additions to the framework of Shi Shen's school. Some constellations of these two schools consisted of stars so faint that they should not be recognized as a result of positional astronomy, but much more of ideological designation and imagination. Apparently there was a certain philosophy behind these faint constellations.

The large number of constellations is one of the characteristics of the Chinese sky. Contrarily, the ancient Sumerian sky contained about 70 constellations[1]. The Greek sky, which more or less adopted the Sumerian tradition, was composed of only 48 constellations at the time of the astronomer Ptolemy. The Chinese constellations were much smaller than the Greek ones; sometimes even one individual star was treated as a constellation. In that case it is more proper to call them asterisms. This character of the Chinese sky is related to another aspect, the naming of the constellations.

[1] Several tablets of Sumerian and ancient Babylonian texts have been found describing the celestial phenomina related to stars; 71 star names were listed in the Mul · Apin text. Papke [1989] made a reconstruction of the Sumerian sky at about 2300 BC based on these descriptions.

With 283 names of constellations and even more names for individual stars in big constellations, the sky is already of interest from a cultural point of view. Most names are not from myths and legends, like in the Sumerian or Greek skies, but from specific material objects related to human society and activity. If we have a brief look at the Chinese star names, we can roughly categorize them as related with:

— Heaven itself: sun, moon and stars as names of constellations;
— Royal court and clan;
— Imperial bureaucracy and administration;
— All kinds of buildings and facilities;
— Military installations; armies and weapons;
— Traffic and transportation;
— Rituals, ceremonies and pictures of social life;
— Philosophical and religious concepts;
— Mythological and legendary figures;
— Administrative provinces and geographical features.

Actually we see an entire cultural complex projected on the sky, characterizing an imperial society!

When Sima Qian compiled his chapter to describe stars he gave it the title *tian guan* 天官, the Celestial Officials. It was explained by later commentators as follows: "The *guan* mean constellations. Constellations in the sky are ranked high and low, like the ranks of officials in human society; hence the name Celestial Officials."[2] Thus the celestial projection was a reflection of human society. The *Tianguan shu* begins with the following passage:

> The Central Palace (*zhong gong* 中宫) is dominated by [the stars of the constellation] *Beiji* 北極 (North Pole Office); the brightest star is the residence of *Taiyi* 太一 (Supreme One); the three stars next to it represent *Sangong* 三公 (Three Excellencies) or, as some people say, princes and sons of imperial concubines. Behind [*Beiji*] are four stars in a curve; the star at

[2]The *Shiji suoyin* 史記索隱 (Hidden Facts Revealed from *Shi ji*), notes and commentaries by Sima Zhen 司馬貞 from the 5th century.

> the end is called *zhen fei* 正妃 (Empress); the other three stars are concubines in the inner palace. Surrounding [the Central Palace] are twelve stars; they are subjugated barbarian lords.

Thus the polar area of the sky was correlated with the terrestrial imperial palace. Zhang Heng even discussed the correlation between the stars and terrestrial things in a more philosophical way. He said that "stars materially originated from the earth below; but their essence was perfected above. They are randomly scattered in the sky, but every one of them has its own distant connections. In the wilderness stars denote articles and objects; at court they denote officials; among people they denote human action."[3]. Thus it was clearly argued that the sky was a reflection of the earthly world and society.

Our purpose in this chapter is to study the Chinese sky in connection with its cultural background. The constellations were designated and named not arbitrarily, but with a philosophy, with reasoning and thinking, with understanding of natural and social phenomena, etc. Thus star names have an "astro-mythological" meaning; their arrangement embodies man's understanding and explanation of nature and society. The sky is a unique document about the historical development of the human mind in Asia. First of all stars were perpetual and almost never changed in human history; moreover the sky was common to all peoples over the entire continent. Thus the sky could reflect ancient cultures, and could provide a common basis for comparison between various cultures. The following questions are of interest:

— What is the significance of the constellation names? How do they correspond to things on the earth below?
— All constellations have their specific locations in the sky; what does each of these locations tell us; why was it located there?
— Constellations have been organized into certain structures; how did they develop and what were the backgrounds?
— Are there any relations between the Chinese sky and the skies of other cultures? Are there any analogies between them?

[3]Zhang Heng, *Ling xian* 靈憲 (Spiritual Constitution), existing fragmentarily in notes by Liu Zhao 劉昭 (fl. 502 - 520) on the *Hou han shu lüli zhi*.

These questions should not be answered one by one and not be treated separately. Together they represent the general question about the cultural background of the Chinese sky. We do not intend to solve all these problems in the present study because of the limits of both our ability and the available material. We will only try to look into them for a better understanding.

5.2 *Star names and the philosophy behind*

5.2.1 Literal meaning and astrological denotation

In most cases the literal meaning of a star name is clear. Only in a few cases are the names strange and their meanings vague due to their antiquity. But the literal translations of star names are not enough for a real understanding of the Chinese sky. For better insight we have to turn to astrological descriptions. Stars played an important role in ancient Chinese astrology; they provided a background against which all celestial phenomena happened. These in turn were related to terrestrial phenomena. Explanations were given by astrologers. Therefore these astrological descriptions became a part of the meaning of star names.

Most descriptions about constellations were in the style of astrological protases. They began with what the constellation denoted or presided over; then they supposed certain celestial phenomena to happen there, after which the astrological interpretation or prediction of events followed. So the astrological protases provided the basic material for the understanding of the meaning of the constellation. Generally speaking all ancient Chinese descriptions of constellations are astrological denotations. However, they also provide some astronomical data such as positions, numbers of stars, brightness, etc.

Different astrological schools had their own designations for the stars which resulted in a diversity of meanings of star names. Nevertheless, all these denotations were usually based on common features or knowledge. Therefore it is possible to deduce the basic interpretation of the name of a constellation by careful analysis and discrimination.

This situation will be illustrated in the following paragraph on the constellation *Sheti* 攝提 which we cite from the *Kaiyuan zhanjing:*

> Master Shi said: "the six stars of *Sheti* are carried by the *Da-*

jiao star on both sides, other names [of it] are Circular Pivot, Celestial Pivot, Facade Dunes, Law Executor, Three Excellencies, Celestial Ax, Celestial Prison, Celestial Shield, Celestial Power, or Celestial Army. They sit on the east and west sides [of *Dajiao*] like two triples; if the triples turn eastwards, the Son of Heaven will be propitious and prosperous; if they turn northwards, his majesty will lose power, a sage will change the social structure, and there will be no peace under heaven, ..."

The *Tianguan shu* said: "The stars of *Sheti* are just in line with the handle of *Bei Dou*, thus indicating seasons, that is why they are called *she ti ge* 攝提格, the starting point of *Sheti*[4]."

Wuxian said: "*Sheti* has [two sets of] three stars formed like footholds of a triple; on both sides of *Dajiao* there is a triple; if they turn towards southwest, Sovereignty will shift to another family."

The *Yuewei setuzheng* 樂緯葉圖徵 [5] said: "*Sheti* is the shield to guard the seat of the emperor."

The *Hezheng tu* 河徵圖 said: "*Sheti* denotes the nine ministers."

The *Luo shu* 雒書 said: "*Sheti* is related to change of policies and shift of power."

The *Dilanxi* 帝覽喜 said: "If the Way of Benevolence is not followed, [the triples] of *Sheti* will lose balance."

The *Hanwenjia* 含文嘉 said: "If the emperor appreciates masters with high respect, [the triples of] *Sheti* will turn to the

[4]It is interesting to note that *Sheti* was connected with three things which were apparently related to each other. Firstly, *Sheti* was the constellation in question. Secondly, *Sheti* was also the synonym of Jupiter, *shui xing* 歲星 (Year Star). Jupiter moved round the sky in about 12 years; ancient Chinese divided the sky into twelve sectors so that the position of Jupiter became a kind of chronicle, hence the name Year Star. Thirdly, according to the Twelve Branch division of the sky, the sector *yin* 寅 (third branch) included *Dajiao*. The beginning of the year (the first month of spring) was indicated by the handle of *Beidou* pointing to the direction of *yin* 寅. This branch *yin* was also called *she ti ge* 攝提格, the starting point of *Sheti*. These indications are by no means accidental, they were all related with the beginning of the year or the twelve year cycle of Jupiter. That was probably also the reason why *xiu Jiao* 角 was the first of the 28 *xiu*.

[5]This and the following books from which the *Kaiyuan zhanjing* was quoting were all "woof" books, *wei shu* 緯書. See footnote on page 23.

right direction; otherwise they turn to the wrong direction."

The *Chunqiu wei* 春秋緯 said: "If *Sheti* is in line with *Beidou* and *Jiao* it helps to reach the populace; if not, a new epoch will start under heaven and the ruler will perish."

Among the amalgam of astrological denotations and protases, we recognize the following messages:

— *Sheti* indicates the beginning of seasons because it is in line with the handle of *Beidou*, the traditional season indicator, and with *Jiao* (α Virginis), the first leading star of the 28 *xiu*. In that sense *Jiao* was the New Year star.

— by indicating seasons *Sheti* becomes related to the shift of sovereignty; in Chinese cosmology the emperor is the representative of Heaven. His task is to issue a new calendar symbolizing the beginning of a new sovereignty[6].

— moreover the bright star *Dajiao* (α Bootes) denotes the royal court or the Big King; thus *Sheti* denotes armies or weapons to guard the court.

This example illustrates how important these astrological descriptions are for the understanding of the significance of a constellation. Famous books like the *Tianguan shu* of Sima Qian, the astronomical chapters of the *Jin shu* 晉書 and the *Sui shu* 隋書, etc. are all astrological descriptions of the stars. "Woof" books also contain this type of material. And they were largely cited in the *Kaiyuan zhanjing* so that the latter became a large collection of astrological descriptions of constellations.

5.2.2 Different ways to name constellations

Now let us look at these constellations. There was always a philosophy behind each constellation, in some cases easy to see, in other cases not. Several ways of naming are shown below.

[6]*Shiji lishu* 史記 • 曆書, the Calendrical Chapter of the Historical Records says: "When the mandate shifts to another family, one must be very careful to make a good start. To change the calendar, to change the colour of costume, to calculate the celestial epoch, that is the way to receive the mandate of Heaven."

Description of shape

It is natural that some constellations received names of articles or objects of daily life simply because they resemble these things in shape. For instance the seven bright stars of Ursa Maior when linked with imaginary lines resemble a dipper with handle, hence the name *Beidou* 北斗 which literally means Northern Dipper. This way of naming stars suggests that some names originate from old times; these constellations usually consist of bright stars and there are no sophisticated philosophies behind them. Other examples are the constellations of:

— *Guansuo* 貫索, the Chain or Loop of Links, which indicates the stars of Corona Borealis;
— *Tianchuan* 天船 the Celestial Boat, the shape resembles a boat and its position is on the Celestial River (the Milky Way);
— *Santai* 三臺, the Three Steps, which are six stars lined up in pairs, forming the three steps of a ladder leading to Heaven;
— *Hushi* 弧矢, the Bow and Arrow, which is clearly visible below Sirius; etc.

Description of the celestial sphere

The designation of some constellations reflects the association with specific parts of the sky. This is typical for some constellations along the ecliptic. The ecliptic is the path for the moving sun, the moon and the planets. Accordingly, one might expect some gates and doorways along the ecliptic. Indeed, there are such constellations as *Tianguan* 天關 (Celestial Pass), *Tianmen* 天門 (Celestial Gate), *Tianjie* 天街 (Celestial Street), *Tianyue* 天鑰 (Celestial Keyhole), *Pingdao* 平道 (Flat Road), etc., all along the ecliptic; they resulted from the attempts to describe the ecliptic. Obviously these celestial roadmarks must have been constructed rather late, earliest after the recognition of the ecliptic[7]. They belong to a systematic way of naming constellations.

[7]The clear conception of the ecliptic was not earlier than the Zhanguo period. It might be from Sima Qian's time (about 100 BC) when the *hun tian* 渾天 (Enveloping Sky) cosmology appeared. Sima Qian mentioned it as *zhong dao* 中道, the Middle Way of the sun and the moon in his *Tianguan shu*. It was as late as the Later Han that astronomers began to talk about the introduction of ecliptical coordinates of celestial objects. See the *Hou han*

Names from religion, myths or legends

The perpetual sky provided a general background on which ancient cultures could be recorded. Religions, myths and legends which form important parts of an ancient culture certainly reflect in the sky by means of the names of constellations. In the Chinese sky there are constellations related to gods and deities like *Huangdi zuo* 黄帝座, *Sidi zuo* 四帝座, *Xuanyuan* 軒轅, *Tianyi* 天一, *Taiyi* 太一, *Tianhuang dadi* 天皇大帝, etc.; there are also legendary or historical personages like *Niulang* 牛郎, *Zhinü* 織女, *Wangliang* 王良, *Zhaofu* 造父, *Xizhong* 奚仲, and *Fuyue* 傅說.[8] These projections might originate from later astrologers, but they were based on very ancient myths.

5.2.3 Cosmological ideas and the development of astrology

The correlation between Heaven and Man

There has been a long lasting belief among the Chinese that a close interaction exists between objects and processes in the physical universe and human affairs; therefore many divination methods were used to predict future misfortune or prosperity or to inquire about the will of the gods. During the Yin dynasty people used to examine cracks on a scorched tortoise shell to ask the opinions of gods or to inquire whether a particular action was blessed by them. In the early Zhou, divination was by means of *gua* 卦 (hexagrams), soothsaying by means of numbers. Later, during the Chunqiu Period, with increasing astronomical knowledge, astrology became one of the most important methods to find out divine favours.

> Astrology is used to arrange the twenty eight lunar lodges in the correct order and to note the progressions of the five planets and of the sun and the moon, so as to record thereby the manifestations of fortune and misfortune. The *Yi jing* 易經 (Book

shu lüli zhi 後漢書 • 律曆志 (Treatise on Harmonics and Calendar of the History of the Later Han). The ecliptical coordinates accompanying the *Shishi xingjing* Star Catalogue were not from the same time as the *qu ji du* 去極度, they were added to it not earlier than the Later Han. See chapter 3.

[8]For the meaning of these star names see appendices.

> of Changes) said: `Looking at signs in the heavens, one thereby ascertains the changes of the seasons.'[9]

Calendrical almanacs were also involved in astrology.

> Almanacs serve to arrange the four seasons in the right succession, to adjust the times of the equinoxes and the solstices, and to note the concordance of the periods of the sun, moon, and the five planets, so thereby to examine the actualities of cold and heat, life and death. These are the arts through which the Sage comes to know the decrees of Heaven.[10]

The belief of interaction between Heaven and Man was deeply rooted in Chinese life. Everywhere symmetry and harmony between heaven and man were emphasized. Heaven and earth were closely related and in complete harmony. China was the center of the flat earth, the "Empire of the Middle", corresponding to the celestial pole, the center of heaven; there the mighty god *Shangdi* 上帝 ruled as did the emperor on earth. The emperor was the Son of Heaven. He received the mandate from *Shangdi* (God of Heaven) to rule over the Empire of the Middle (*zhong guo* 中國) and had to maintain the harmony between heaven and earth by his actions in following precisely the rituals and prescripts of his forefathers. The basis of all rituals and actions by the emperor and his officials was the annual calendar announced by the emperor in the Bright Hall *Mingtang* 明堂 in the first month of the new year. Disorder on the part of one relation brought disorder into all others. That meant that not only irregularities in heaven caused calamities on earth but also evil actions of man caused disturbances in nature and in heaven. Eclipses and comets could be interpreted as indications that the emperor or the officials sinned or governed badly. This belief was a fertile soil in which astrology could grow. Phenomena happening in heaven were the most direct manifestations of the interaction between Heaven and Man and a warning to the emperor.

---

[9]*Han shu yiwen zhi* 漢書・藝文志 (Treatise on Bibliography of the History of the Former Han).

[10]*ibid.*

The Yin-yang *philosophy and the Five Elements (Phases)*

The philosophy of the interaction between Heaven and Man was dramatically developed during the Former Han when the concepts of the *yin-yang* and the Five Elements were interwoven into it. The *yin* and *yang* were conceived of as two complementary principles or forces, of which the *yin* represented feminity, darkness, cold, passivity, etc., while the *yang* represented masculinity, light, warmth, activity, etc. According to the *Yin-yang* school which started in the third century BC[11], all natural phenomena resulted from the interplay of these two forces. The Five Elements, earth, wood, metal, fire and water, were regarded as different natures or forces, eventually called "phases", rather than as actual states of matter. The Elements succeeded (*ke* 尅) and produced (*sheng* 生) each other in fixed sequences. When an object was believed to embody a certain Element (Phase), it would interact with other objects (Elements) according to the rules of interaction between the Elements (Phases). Thus all phenomena, natural or social, could be explained by the fundamental relations between the Five Elements (Phases). For example — spring corresponded with wood, summer corresponded with fire, but as fire always succeeded wood, the summer had to follow spring.

On this simple and intuitive basis a cosmological system was developed, based on correlations made between

— the Five Elements: earth, wood, metal, fire and water;
— the four compass points plus the center as the fifth point;
— the four seasons plus the last month of summer;
— the five planets Mercury (water), Venus (metal), Mars (fire), Jupiter (wood), and Saturn (earth);
— the five notes of the musical scale: *gong* 宮, *shang* 商, *jiao* 角, *zhi* 徵, and *lü* 呂;
— the five colours: green, red, white, black and yellow;
— the five tastes: acid, sweet, bitter, peppery, and salty;
— also the twelve months, the twelve pitch-pipes, the ten heavenly stems, the twelve earthly branches, etc.

[11] The progenitor of this school was Zhou Yan 鄒衍 who lived in the third century BC in the State of Qi 齊 (present Shandong province). See the *Shi ji*, ch. 74.

The universe was conceived of as an ultimate integration of natural and social forces.

During the Former Han, Dong Zhongshu 董仲舒 started to expound Confucian classics in terms of the cosmology of the *Yin-yang* school. The interaction between heaven and man was emphasized more in his philosophy than ever before. All his writings served to elucidate the meaning of the classics, incorporating the ideology of the *Yin-yang* school. This was done with deliberation in his *Chunqiu fanlu* 春秋繁露 (Luxurant Dew of the Spring and Autumn Annals). After he was consulted by the Emperor Wu on three fundamental questions about how to maintain the harmony of the universe[12], his newly expounded Confucianism got official recognition. This received great support among the Han scholarship. The close interrelationship between the "Way of Heaven" and human affairs was taken for granted, and it became common for confucianists to interpret unusual natural phenomena in terms of celestial vistas or prodigies.

Astrology acquired the character of being "logical" and "convincing" when the basic cosmological ideas were incorporated to interpret celestial phenomena in correlation with human affairs. For example,

— the five planets were correlated to the Five Elements
— the Sun and the Moon were related to the *yang* and the *yin* respectively
— the part of the sky North of the ecliptic was *yang*, the Southern part was *yin*
— certain constellations were assigned to certain Elements.

Thus celestial phenomena could be easily seen as a result of the interplay of the forces of *yin* and *yang* on the Five Elements. Astrological interpretations could be readily derived. Astrology experienced a great progress during the Former Han; it became more systematic and sophisticated.

The field division theory and the astrological scheme of the sky

Astrology dealt with phenomena happening in the sky and their relations with terrestrial affairs. For the optimal functioning of astrology it was necessary to establish a fixed framework, a real coordinate system in the sky

[12]His answer to these questions was known as *tian ren dui ce* 天人對策, the discourses on the relation between Heaven and Man. See the *Han shu*, ch.56.

Table 5.1: The "Field Division" (*fen ye* 分野) of the *xiu*.

Lunar Lodges, *xiu*	States	Lunar Lodges, *xiu*	States
Jiao, Kang, Di	Han	*Kui, Lou, Wei*	Lu
Fang, Xin	Song	*Mao, Bi*	Zhao
Wei, Ji	Yan	*Zui, Shen*	Wei
Dou, Niu	Wu and Yue	*Jing, Gui*	Qin
Nu, Xu, Wei	Qi	*Liu, Xin, Zhang*	Zhou
Shi, Bi	Wei	*Yi, Zhen*	Chu

in correlation with the social structure on the earth. Unusual phenomena could always be regarded as deviations from normal structures; thus they would have unusual implications. This fixed framework was established by a correlation between fixed stars and terrestrial objects, specifically by designating constellations. This gave impetus to the construction of more constellations in the sky. The names of these constellations became the astrological framework.

Following this kind of thinking, a practice developed to correlate the sectors of the twenty eight *xiu* with the regional states. This method was called the divination of the *fen ye* 分野 ("Field Division"). It might have originated from the Zhanguo period, but it became more sophisticated during the Han[13]. According to the geographical chapter of *Han shu*, the *fen ye* correlation between the regional states and the *xiu* was as shown in the Table 5.1.

What happened in the sky could be correlated with events in various states; astrology became of vital importance for the central government. Court astrologers examined what happened in the sky; their information to the central court was used by the emperor for actual decisions about human welfare in remote states, and thus exerted a menacing power.

The *fen ye* method was developed by the political astrologer Liu Xin 劉歆 (53 BC - AD 23) and was widely used.

The correlation between the heavens and the earthly society according to the *fen ye* of the 28 *xiu* was also extended to other constellations. These

[13]See Li Yong 李勇 [1992].

developments happened parallel to and in agreement with Dong Zhongshu's philosophy about the correlation between heaven and man. Following Dong Zhongshu's *Chunqiu fanlu*, there appeared a lot of so-called "woof" books or *wei shu* which were a kind of commentaries on the six Confucian Classics, namely, on the Book of Odes (*Shi* 詩), the Book of Documents (*Shu* 書), the Book of Rituals (*Li* 禮), Music (*Yue* 樂), the Book of Changes (*Yi* 易) and the Spring and Autumn Annals (*Chun qiu* 春秋). The "woof" books were meant to combine with the classical "warp" books to form the general body of knowledge about nature and society. They were all texts from the Han time or later, written for the purpose of interpreting the Classics in terms of the new *yin-yang* ideology. In the "woof" books there were many descriptions about denotations of constellations. These denotations formed an important part of the astrological framework in the Chinese sky. As we have seen from quotations from *Kaiyuan zhanjing*, the texts of *wei shu* contributed a lot to the astrological meanings of various constellations.

All astrological descriptions about the sky from the middle of the Former Han onward were aimed at the construction of a correlating frame between the celestial background and the terrestrial society. It was inevitably influenced by the *Yin-yang* and Five Element ideology as accepted during the Han. This does not mean that the astrological meanings of constellations have all been invented by Han astrologers. Ancient knowledge about the sky was certainly inherited and new astrological denotations were based on ancient traditions; otherwise they would not have been accepted because of inconsistency with ancient knowledge. By means of the astrological structure from the Han time, we can still go further back to investigate the origin and meaning of constellations.

5.3 The relation of the sky to the seasons

The appearance of the sky changes with the seasons because the sun moves around the sky during the year. Stars which are opposite to the sun on the celestial sphere are visible during the whole night. Stars to the east of the sun are visible after sunset; the closer they are to the sun, the shorter the time they can be seen before setting in the west. Stars to the west of the sun are visible before sunrise; the closer they are to the sun, the shorter the time they can be seen before disappearing in sunlight. Therefore, stars can be used to determine the position of the sun and thus to determine the season. This is the astronomical reason why the sky is related to the seasons. This knowledge was already applied by man since remote ancient times. A well known example is the ancient Egyptians observing the heliacal rising of the star Sirius in order to know when the flood by the river Nile would start. According to the above mentioned *Yaodian* (Canon of Yao), the culmination of one of the four cardinal asterisms at dusk was used to indicate a corresponding season by the Chinese at about 2300 BC.

But we may look at this relation the other way around: the meaning of constellations has certainly been influenced by things happening during the seasons with which they corresponded. Thus the relation between sky and season is one of the keys to understanding the positions, names and shapes of the constellations.

As we have seen above the Chinese sky is a projection of terrestrial society. Constellations have emerged gradually during a long series of annually recurring events, of atmospherical or terrestrial phenomena as well as of self-repetitive social happenings like offering rituals or New Year ceremonies at court.

The seasons regulated Man's Life:

— Seeding and harvesting should happen at the right time of the year to prevent starvation.

— Military expeditions should happen after the gathering of the harvest but before midwinter.

— Sacrifices and ceremonies to honour Hou Ji 后稷 (God of Grains) and Huangdi 黄帝 should happen shortly after the full Moon had passed the Cardinal Points in the Sky. These were the traditional Spring, Summer, Autumn and Winter ceremonies.

To live in that way meant to obey the Rules of Heaven and to be in Harmony with the Cosmos. Shortly after New Year, the Emperor (*tian zi* 天子, Son of Heaven) presented his plans for the year in *Mingtang*. These plans should be in strict conformity with the Celestial Calendar to prevent disasters. Therefore the Celestial Calendar had to be an exact reproduction of terrestrial life. The position of Sun and Moon on this calendar told the Royal Astrologers what should happen and what was coming. There was a fixed calendar for rituals and ceremonies which were symbolic acts for natural and social life. In doing so, the Way of Heaven was followed and

Universal Harmony would be achieved.

5.3.1 Astronomical Lunar and Solar Calendars

Two astronomical concepts are essential to understand the correlation between stars (or constellations) and the seasons: the Lunar Calendar and the Solar Calendar.

By definition the Lunar Calendar indicates the position of the full moon at a certain time of the year. Constellations near that position correspond to that season. This was probably the most commonly used way in ancient times, because that part of the sky where the full moon would appear was in opposition to the sun and was visible during the whole night. The invention of the lunar lodges is obvious evidence of this practice.

Similarly, the Solar Calendar meant that the position of the sun was used to indicate the season. Constellations near that position should be related to the corresponding season. The problem was that the constellations in question were not visible because of the sunlight. Then some methods must be found to determine the position of the sun among stars. In ancient China, heliacal observations were not well developed as in the ancient Middle East. The solar position among the constellations was actually determined by the observation of the position of the full moon or by the observation of stars culminating at dusk, dawn or midnight. To do so, the complete picture of the sky and a sophisticated knowledge of the movements of sun and moon were needed.

The determination of the solar position among the stars was a later development of astronomy in China. The position of the sun among the *xiu* was first systematically registered in the *Yue ling* 月令 (Monthly Ordinances) which was completed during the Qin. Actually the first sentence for each month of the *Yue ling* is a description of the position of the sun among the lunar lodges (*xiu*). In an earlier book of this type, the *Xia xiao zheng* 夏小正 (Small Calendar of the Xia Dynasty), however, the position of the sun was not mentioned at all; heliacal and meridian observations of stars were mixed up in the text. Apparently the *Yue ling* incorporated the most advanced calendrical and astronomical knowledge around the Qin time.

Then the question is: for a given part of the Chinese sky which Calendar, or which method of correlation should be applied? This is actually a question about how to correlate a particular star or constellation with the season. It is to be discussed later when we examine the Chinese sky in detail. But here, a general explanation seems to be neccessary.

The *Lunar Calendar*, as we will use it, essentially deals with the opposition point of the sun (anti-sun) on the ecliptic. There is a difference of half a year in the seasons relative to the Solar Calendar. At the position of the anti-sun, the Full Moon will always appear.

In ancient times, probably all through the Yin and the early Zhou, a primitive Lunar Calendar was in use. The appearance of the Full Moon in or near a certain constellation indicated the time of the year. The Full Moon between *xiu Lou* 婁 and *xiu Mao* 昴 meant Mid-autumn. The Lunar Calendar might be applied to this segment of the sky.

Starting from here along the ecliptic, there followed the *xiu Bi* 畢, *Zui* 觜, *Shen* 參, *Jing* 井 and *Gui* 鬼, and the constellations *Xuanyuan* 軒轅 and *Taiwei* 太微, and the *xiu Jiao* 角. This half sector of the sky was apparently based on the Lunar Calendar[14]. That means that the months from the Autumn equinox to the Spring equinox were related to this calendar. It was the winter sky.

The *Solar Calendar*, as we will use it, indicated the celestial position of the Sun during the year. When the Sun was near *xiu Jiao*, α Virginis, then it was Mid-autumn (22nd September) during the Han; and when it approached *xiu Niu* 牛, β Capricornus, it was Mid-winter. Farmers' calendars, *xia li* 夏曆, followed essentially the same system. Starting from *xiu Jiao* 角 along the ecliptic, followed *xiu Kang* 亢, *Di* 氐, *Fang* 房, *Xin* 心, *Wei* 尾, *Ji* 箕, *Dou* 斗, *Niu* 牛, *Nü* 女, *Xu* 虛, *Wei* 危, *Shi* 室, *Bi* 壁 and *Kui* 奎. This half sector of the sky might be applied to the Solar Calendar[15]. It was again related to the months from mid-Autumn to mid-Spring.

Still, which calendar is applicable to a particular constellation is a rather complicated problem. Celestial context, astrological meaning, and seasonal circumstances should be taken into consideration to decide the right calendar. From an observational point of view, there is a difference between the Lunar Calendar and the Solar Calendar: stars at the position of the Full Moon are visible whereas stars at the position of the sun are not visible.

[14]Obvious evidence was that the Moon Star *Yue* 月 of the *Gan Shi* was indicated in the *xiu Mao* 昴.

[15]Similarly, the Sun Star *Ri* 日 of *Gan Shi* was indicated near *Fang* 房 *xiu*.

That means constellations which are connected to the Solar Calendar must have been constructed by professional astrologers who had a picture of the entire sky in mind and had adequate knowledge about the motion of the sun. Therefore, the Solar Calendar is most applicable to those constellations of later origin, namely, of Han times.

The lunar lodge system provided celestial reference points for the measurement of lunar and solar positions. Thus it was essential for calendar making. It is remarkable that the Chinese *xiu* constellations do not always follow the ecliptic. In the Chinese system, the *xiu Xing* 星, *Zhang* 張, *Yi* 翼 and *Zhen* 軫 apparently follow the equator. It seems that these *xiu* were deliberately chosen to indicate the equator, unlike the Indian *yogathara* and Arabian *anwa* which follow the ecliptic here[16]. Because of this essential difference between the Chinese and Indian lunar lodge systems and the mutual influences we see in this section of the sky aspects of both Lunar (winter) and Solar (summer) Calendar.

Because the Chinese sky was not constructed during one single period, overlaps of Calendars also happened in other parts of the sky. Explaining the names of constellations, one must be careful to choose the right calendar. The general criterion has been indicated above, but one should always consider celestial and terrestrial contexts in order to make a definite choice of the calendar.

5.3.2 The effect of precession

Due to the precession of the celestial pole and the equator, the equinoxes and solstices move one degree westward along the ecliptic in about seventy years. The precession phenomenon was recognized in China in 330 AD by

[16]An equatorial coordinate system was adopted by the Han astronomers, about 100 BC. Apparently they have chosen the positions of the *ju xing* along the equator for the "winter-half of the sky", whereas the *ju xing* for the "summer half" have a tendency to follow the ecliptic. The winter half is described on a lunar calendar and the summer half on a solar calendar, both with exceptions.

In India about 100 BC only ecliptical coordinates were used, according to Burgess [1935] and Pingree [1978]. The Arabs following the example of the Indians after 800 AD did the same for their system of *Manazils*; see Kunitzsch [1991].

Yu Xi 虞喜[17]. But already during the Han it had been noticed that the winter solstice shifted among the *xiu*[18]. The four cardinal points of the Yao time seem to be well known to astronomers of later times. Indeed the precession was discovered by comparing the four cardinal points from the *Yaodian* with Yu Xi's own observations. Actually, the Yao time was supposed to be at about 2300 BC by astronomers of Yu Xi's time[19], the same as the result of modern research. So there was apparently an astronomical tradition by this early date. This tradition was probably known to Han astronomers. Studying the origin of constellations one should always keep this date in mind, because ancient memories could be reflected even in constellations constructed by Han astrologers.

Precession caused a shift in the correlation of stars with seasons. There were about 2200 years from the time of Yao to the Han time corresponding with a shift of about one month. This shift should be taken into account when we try to connect ancient constellations like *Bi, Shen* and *Wuche* with the seasons. The general picture is from the Han time, but now and then more ancient dates should be taken into consideration.

---

[17]Yu Xi discovered a constant called *shui cha* 歲差which is the difference between the lengths of the sidereal and tropical year. This he explains with the annual westward shift of the winter solstice along the ecliptic.

[18]During the early Former Han, the ancient *si fen* 四分 Calendar was used, in which the winter solstice was put in *xiu Niu* 牛. In 25 AD Liu Xin 劉歆 observed the Mid-winter Point at 4.5 degrees West of *xiu Niu*, and in 150 AD, Li Fan 李梵 measured 6.5 degrees West of *xiu Niu* as described in the *hou si fen* 後四分 Calendar. See the *Hou han shu lüli zhi*.

[19]The astronomer of the Southern Dynasties He Chentian 何承天 (370 - 447) said the time of Yao was about 2700 years before his own time (443 AD). See the *Song shu* 宋書 (Book of the Song Dynasty (420-479)).

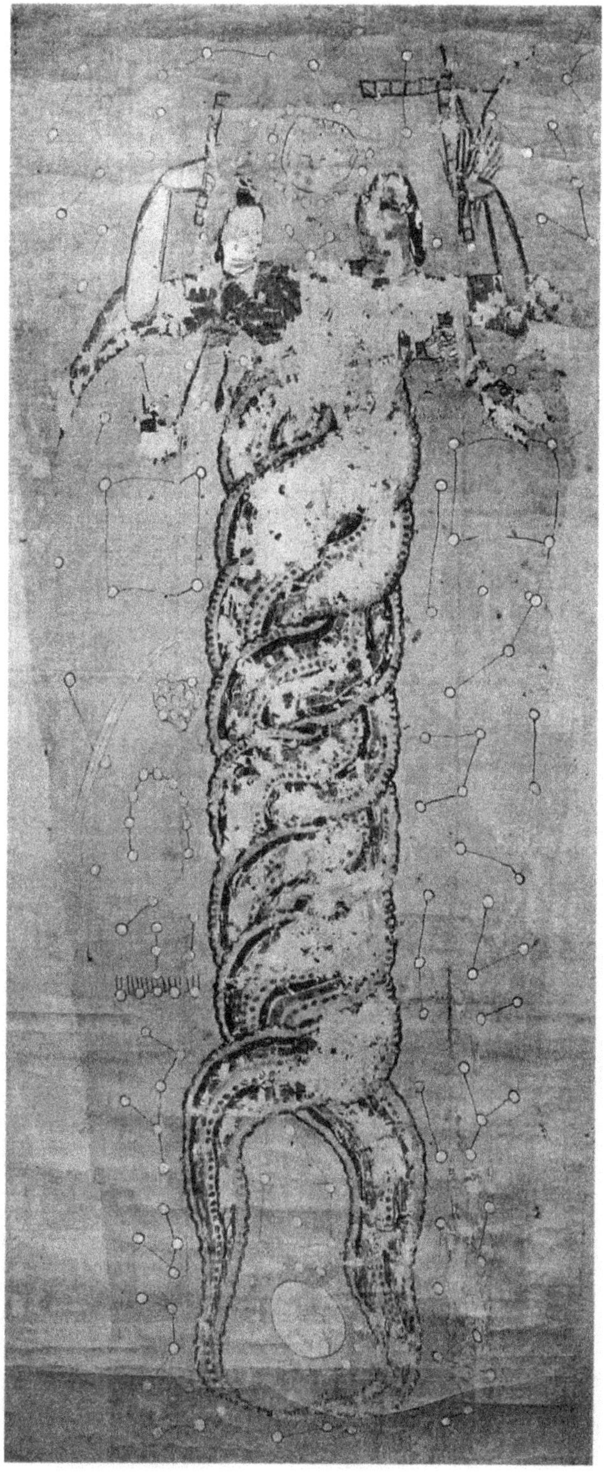

Silk painting unearthed at Gaochang (Turfan), dating from about 500 AD. It portrays Fu Xi and Nü Wa, the creators and inventors in Chinese mythology. Surrounding them is the cosmos, represented with the Sun, the Moon, and constellations. Most of the constellations are lunar lodges (*xiu*). Photo extracted from the *Zhongguo gudai tianwen wenwu tuji* (Wenwu Press, Beijing; 1980).

CHAPTER SIX

MAIN STRUCTURES IN THE SKY AND THEIR MEANINGS

6.1 *The twenty eight* xiu *and the four Images*

As we have seen in the short historical survey of the Chinese constellations[1], the twenty eight lunar lodges, the *xiu* are among the earliest constellations in the Chinese sky. They also formed its basic and important framework. At about the Han time these *xiu* were divided into four quarters and were related to four compass directions with four Animal Images, the *xiang* 象 as shown in Table 6.1.

The *xiu* more or less followed the ecliptic, forming a system of reference points to register the movement of the sun, moon and five planets and the relative positions of other stars; thus they were essential for calendar making. That not only China but also India and the Arabic world had the 28 lunar lodges has already evoked many hypotheses about the origin and the transmission of the system. Although opinions diverge as to where it was first invented, agreement has been reached that these systems must have had the same origin. Thus for the astro-mythological study of the sky the meaning of the 28 *xiu* is of primary significance. Firstly, because they are very old, they may convey a very ancient picture of culture; secondly, because there are similar systems in other cultures, a comparison might eventually throw light on the ultimate origin of the system.

The fact that most of the oldest *xiu* names consist of only one Chinese character (this already distinguishes them from the constellations desig-

---

[1]See Chapter 2.

Table 6.1: The *xiu* and the Four Images.

East seven *xiu*	South seven *xiu*	West seven *xiu*	North seven *xiu*
Blue Dragon	Red Bird	White Tiger	Black Warrior
Jiao 角	Jing 井	Kui 奎	Dou 斗
Kang 亢	Gui 鬼	Lou 婁	Niu 牛
Di 氐	Liu 柳	Wei 胃	Nu 女
Fang 房	Xing 星	Mao 昴	Xu 虛
Xin 心	Zhang 張	Bi 畢	Wei 危
Wei 尾	Yi 翼	Zui 觜	Shi 室
Ji 箕	Zhen 軫	Shen 參	Bi 壁

nated later) makes the understanding of the meaning of these names more difficult. The literal meaning could be vague and ambiguous. The antiquity of these *xiu* and the important part they played in the sky both contributed to more astrological explanations, either derived from ancient tradition or from fantasy. Sometimes a better understanding of their meaning can be obtained by reading the celestial context of these *xiu*.

*The Image of the Green Dragon (*cang long*).* The seven eastern *xiu* were related to the Green Dragon (*cang long* 蒼龍 or *qing long* 青龍)[2]. Indeed, the first six of them were exactly a description of the body of a dragon, from horns to tail. The *Tianguan shu* describes more explicitely how they represented the body of a dragon. The recently discovered star chart on the ceiling of a tomb from the Former Han in Xi'an Jiaotong University also shows a picture of a dragon based on six *xiu*. See Figure 6.1. Even more interesting is that the dragon's heart, *xiu Xin* 心, α Sco., was painted as the only red star on the chart, to indicate the Fire star *Huo*, one of the cardinal asterisms of the Canon of Yao. In the Yin time the appearance of *Huo* above the eastern horizon at dusk indicated the arrival of spring, the beginning of

[2]Some people translated *qing long* as Blue Dragon because the Chinese character *qing* 青 can also mean Blue. But as this Image was related with spring, we think Green is better because of its association with spring.

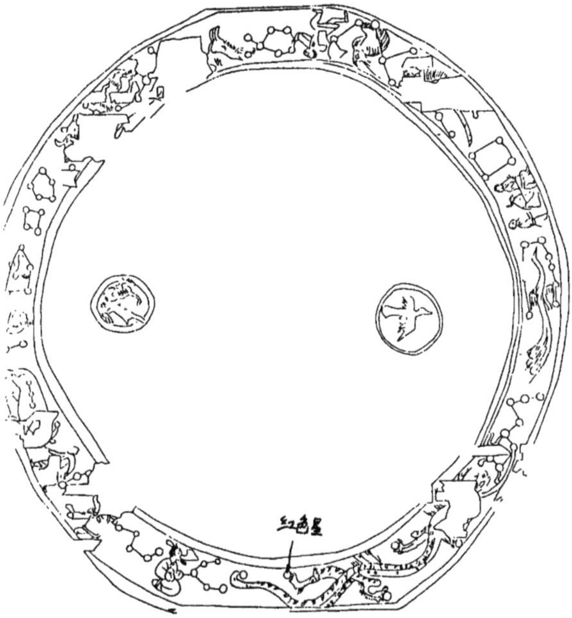

Figure 6.1: Star chart on the ceiling of a Han tomb discovered in Xi'an Jiaotong University. The twenty eight *xiu* were presented in the circular belt. Animal images and other figures were painted to represent the meaning of the *xiu*. One star on the body of the dragon was specially painted red, which was the Fire star.

the year[3]. At that time a big ceremony took place, which was called " *yue* 雩", to pray for timely rains for agriculture during the coming season. This ceremony happened in the middle of the spring when "the dragon appeared above the eastern horizon (*long xian er yue* 龍見而雩)."[4]

It clearly demonstrates that during the Yin the *xiu* alongside the Fire Star were already related to the dragon. The association of the seven eastern *xiu* with the Green Dragon was apparently very old. Indeed, in the recently discovered Puyang grave mosaicked pictures of a dragon and a tiger are respectively on the east and the west side of the buried chief[5]. According

---

[3]Pang Pu [1990] has suggested that a calendar based on the observation of the Fire star was used by the Yin people.

[4]See the *Zuo zhuan* 左傳, the fifth year of Duke Huan 桓 (715 BC).

[5]See Feng Shi 馮時, "Zhongguo zhaoqi xingxiangtu yanjiu" 中國早期星象圖研究 (A study of early star charts in China), *Zhiran kexueshi yanjiu* 自然科學史研究 (Studies in

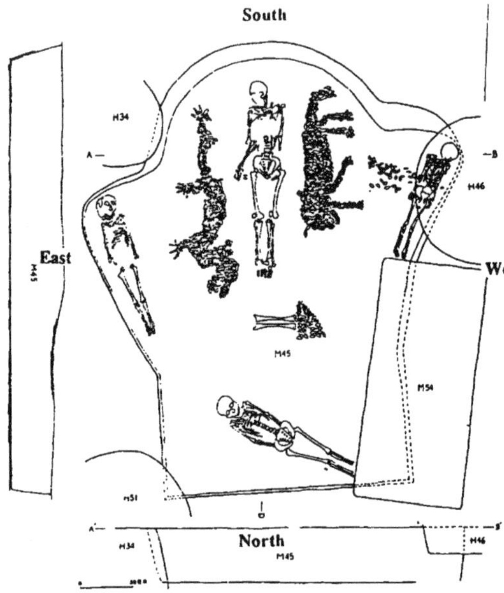

Figure 6.2: Picture of Puyang 濮陽 tomb excavated in 1988 near Puyang, Henan 河南. The mosaicked images from river shells represented the Dragon and the Tiger. To the north, the two leg-bones with a pile of shells attached to them probably indicated the constellation of *Beidou*. The age of this tomb is 3000 BC.

to carbon 14 dating, this grave dates from about 3000 BC in the neolithic times[6]. See Figure 6.2.

*The Image of the Red Bird (*zhu que*).* The southern *xiu* were related to the image of the Red Bird (*zhu que* 朱雀). According to the *Tianguan shu* four of them, namely *Liu, Xing, Zhang* and *Yi*, indicated parts of a bird. This bird was also shown at exactly the same position on the star chart of the Han tomb at Xi'an Jiaotong University. Like the Dragon, this Bird was also an ancient product of imagination called *feng huang* 鳳凰 the Phoenix. The Bird star was mentioned in the Canon of Yao. In the oracle-bone inscrip-

the History of Natural Science), 2 (1990).

[6]C-14 dating was done independently in Beijing and in the Center for Isotope Research of the University of Groningen by Dr. J. van der Plicht.

tions the Bird star has also been registered[7].

Later development of symbolic ideas associated the Dragon with masculine aspects of the world, hence with the Emperor, whereas the Phoenix became more or less synonymous with feminine aspects, hence with the Queen. Thus a "harmonious" picture of a Dragon and a Phoenix symbolized the harmony of the universe.

*The Image of the White Tiger (*bai hu*).* The western *xiu* were related to the image of the White Tiger (*bai hu* 白虎). According to the *Tianguan shu*, only *Shen* 参 *xiu* was described as the body of a tiger, with *Zui* 觜 *xiu* as its head. The names of other *xiu* had no obvious connection with it. This simple fact demonstrates that the correlation between the *xiu* and the Four Images has been built up gradually. The final picture of the White Tiger covering six or seven *xiu* seems to be from the Han time.

The Image of the Black Warrior (xuan wu). The fourth animal image was called *xuan wu* 玄武 (Black Warrior). On bas-relieves from Han times it was shown as an image of a turtle. Because the big river turtle *Bie* 鱉 was one of the sacred animals the shell of which was used for divination[8], and because its shell symbolized a suit of armour, the turtle was definitely connected to the Black Warrior. Sometimes this image was called *ling gui* 靈龜, the Spiritual Turtle. Both in the *Tianguan shu* and on the chart of the Han tomb only *Xu* 虛 *xiu* and *Wei* 危 *xiu* (Aquarius sector of the Zodiac) were connected to the Image of the Turtle. This turtle, strangely enough, had a giant snake entangled around its body. This snake probably symbolized the Dragon. We want to mention here that the seventh *xiu Ji* 箕 of *cang long* contains the ancient *Bie* constellation (Corona Australis), exactly in the position to be entangled by the tail of *cang long*. The legendary Fu Xi 伏羲 discovered the numerological picture of the *Luo shu* 雒書 (the "Luo River Writing")[9] on the back of the River Turtle (*Bie*) . So it might be said that both the *xiu* and the Four Images had their roots in very ancient

[7]Needham, *Science and Civilization in China*, section 20, p.244.

[8]This type of divination might originate from the Yin time. It was frequently recorded in the *Zuo zhuan*. See Fung Yu-lan, *A History of Chinese Philosophy* (translated by Derk Bodde), Delhi edition (1994), Vol. I, p.28.

[9]Needham, Science and Civilization in China, section 19, p.56.

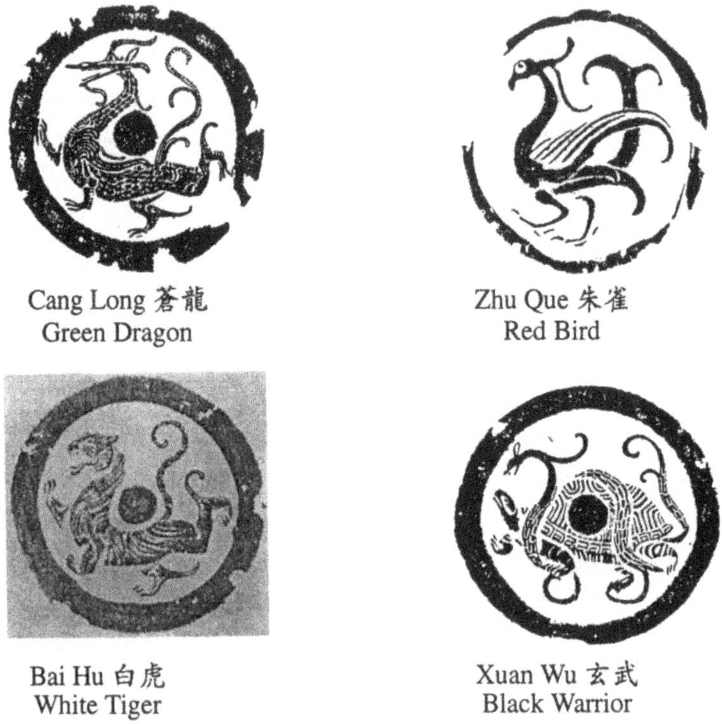

Figure 6.3: The Four Animal Images as shown on eave tiles from the Former Han.

times, but the systematic 28 *xiu* and their connections with the Four Images were developed later. Actually one can easily see the influence of the Five Element theory from the colours of the Four Images which were assigned according to the Five Element scheme of cosmology. Pictures of the Four Images were found on clay tiles from the Former Han. See Figure 6.3.

Apparently the constellations in the Chinese sky were from two stages of development. The *xiu* and several conspicuous constellations were from the first stage which ended with the complete *xiu* system. The second stage was characteristic of an idealistic reconstruction of human society in the sky, which was completed during the Han. Naturally these reconstructions were based on the constellations of the first stage. Thus the *xiu* became the basis for all later constructions. That means that later constellations were a sort of extension or explanation of their related *xiu* or that they described certain ancient pictures which had been related to the *xiu*. Some obvious examples can be found among the constellations of all three astronomical schools.

For example, with respect to *xiu Jing* 井, *Jing* means a water well; thus *Tianguan shu* says "the Eastern *Jing* is related to water affairs." We find around *xiu Jing* the *Shi Shi* constellations *Nanhe* 南河 (South River), *Beihe* 北河 (North River), *Jishui* 積水 (Stored Water), and *Shuiwei* 水位 (Water Level), and the *Gan Shi* constellations *Sidu* 四瀆 (Four Rivers) and *Shuifu* 水府 (Water Palace). If one notices that *xiu Jing* is located just to the east of the Celestial River (the Milky Way), then one can easily understand this general picture of water and rivers in *Jing xiu*.

6.2 Five Palaces

Tianguan shu organized the constellations into five palaces, the four cardinal ones plus the central one. The Central Palace (*zhong gong* 中宮) is the above mentioned Purple Palace (*zi gong* 紫宮) surrounding the Celestial North Pole. The four cardinal palaces might have been intended to be related with the four animal images and to include the four-times-seven *xiu*, but they were described in quite a different way as follows:

> The Eastern Palace (*dong gong* 東宮) (is symbolized by the) Green Dragon, (and its *xiu*) are *Fang* and *Xin;*
>
> The Southern Palace (*nan gong* 南宮) (is symbolized by the) Red Bird, (and its stars) are *Quan* 權 and *Heng* 衡;
>
> The Western Palace (*xi gong* 西宮) (is symbolized by) *Xianchi* 咸池 (the Harmony Pond), (and its stars) are called *Tianwuhuang* 天五潢 (Five Celestial Ponds) or the wagon house of the Five *Di*;
>
> The Northern Palace (*bei gong* 北宮) (is symbolized by the) Black Warrior, (and its *xiu*) are *Xu* 虛 and *Wei* 危.

According to the text, the western palace is related to *Xianchi* (in Auriga) rather than to the White Tiger, as one might expect from the system. Either the characters *bai hu* 白虎 have been lost from *Tianguan shu* or it suggests a different system of Four Images. More striking is that the southern and western palaces are not even related with the *xiu* but with constellations close to the ecliptic. In the Southern Palace *Quan* and *Heng* represent the constellations *Xuanyuan* 軒轅 and *Taiwei* 太微 (all in Leo); and in the Western Palace, the constellation obviously is *Wuche* 五車 (Auriga). Thus all four palaces seem to follow the ecliptic. This may indicate that they were originally related with the motion of the sun or, mythologically speaking, they were the palaces of the Solar God.

What was the Solar God in Chinese mythology? At first sight it seems that the solar god was not very important in Chinese mythology. But this is only because of the systematization of myths by Confucianism and the *yin-yang* philosophies[10]. In fact, the stories about the Five *Di* were most probably the stories about the Solar God; the prototype Huangdi was the Solar God, whereas the Four *Di* were the four gods of the seasons.

Here a little more explanation may be needed.

Up to the Han time, the *Yin-yang* and Five Elements school had already established a world view in which everything was correlated with the Five Elements. Mythology was also incorporated into this general cosmology. This world view was systematically described in some Han books such as the *Huainan zi*. A description about the Five *Di* (and their associated spirits, assistants, etc.) is summarized in Figure 6.4.

In this world view, the Four *Di* were put in the four seasonal palaces at the four cardinal points, which still reflect the aspects of the Solar God. Huangdi was put in the center to be related with the Element of Earth, the original meaning being lost because of the need of systematization. But in ancient literature there were enough descriptions to convince us that Huangdi is the Solar God. For example, it is said in the *Shi zi* 尸子 (Book of Master Shi)[11] that "Huangdi has four faces (*Huangdi shi mian* 黄帝四面);" this can only mean Huangdi was related to four seasons and four directions. Huangdi was also named as Xuan Yuan, because he was born at a

[10]In the first chapter of the *Shi ji* we find the biographies of the Five *Di* 帝, in which Huangdi and the following four *Di*, Zhuan Xu 顓頊, Ku 嚳, Yao 堯 and Shun 舜 were all treated as ancestor kings. This was apparently the historical Confucian version of the Five *Di*. For the *Yin-yang* school's version, see below.

[11]This book was attributed to Shi Jiao 尸佼 of about 4th century BC, but it was probably from a much later time, about 3rd century.

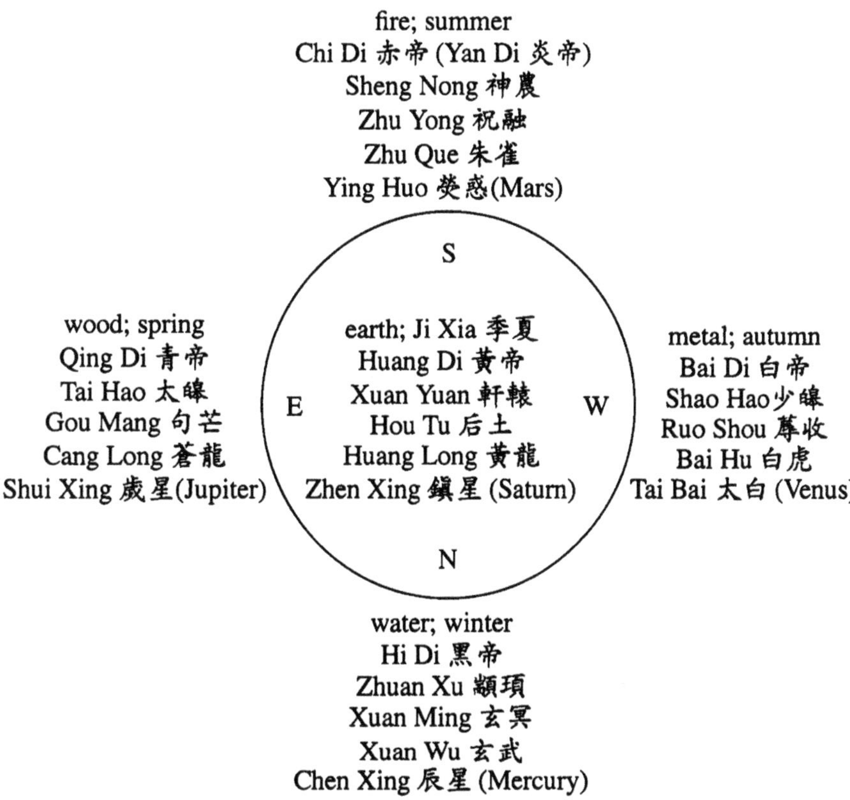

Figure 6.4: The *Yin-yang* school's correlations of Elements, seasons, Five *Di*, names of *Di*, assistants of *Di*, Animal Images, and planets. *Huang long* 黃龍 (Yellow Dragon) was the Image for the Center, which corresponded with *ji xia* 季夏, the third month of the summer. This diagram is based on the *Huainan zi* (Book of Prince Huainan).

dune named Xuan Yuan[12]. Keeping this and the above diagram in mind we read a paragraph from the *Chu ci* 楚辭 (Elegies of Chu):

> (The mount of) Xuan Yuan 軒轅 is unreachable, thus I'll go and play with Wang Qiao 王喬. ... I will pass (the place of) Gou Mang 句芒, leave (the *Di* of) Tai Hao 太皞, and go westwards. ... I meet Ruo Shou 蓐收 together with the Emperor of West, ... I drive directly towards the Yan Di 炎帝, and I will be going to the southern country. I am advised by Zhu Yong 祝融 to go back, ... thus I follow Zhuan Xu 顓頊 to the ever-frozen land, where I pass the territory of Xuan Ming 玄冥 and from there I look back to where I began my journey.

By comparing this text with the above diagram we can see that this poem apparently reflects the annual journey of the Solar God. He moves across the sky and produces four seasons.

In the Chinese genealogy of gods, the Solar God has never been regarded as the Supreme God of the Universe. This High God was the God of Heaven, *Tian* 天, *Di* 帝, or *Shangdi* 上帝. The Solar God was at the secondary or the "atmospherical" level. So all meteorological phenomena were believed to be related to the Solar God; Huangdi was also regarded as the god of thunder and lightning. Then the problem was where to put Huangdi in the sky? He was not as high as the Supreme God of Heaven to be put in the polar area. He therefore needed a place on the ecliptic. Some people would say that Huangdi was seated in the Court of *Taiwei* (see below). In the sky there is an ancient constellation called *Xuanyuan*, which is mentioned in the *Tianguan shu* as *Quan*. The explanation in the *Shiji zhengyi* 史記正義 (Notes and Comments on the *Shi ji*)[13] goes like this:

> *Xuanyuan* 軒轅 consists of seventeen stars to the north of seven stars of (*xiu Xing*). It is the body of the Yellow Dragon (*huang long* 黃龍), the god of lightning and thunder. It also symbolizes the Queen's Palace. When the *yin* and *yang* forces meet, acting violently they produce thunder and lightning; acting gently they

[12]See the *Shi ji*, ch.1, *wudi benji* 五帝本紀 (Biographies of the Five *Di*).

[13]This book was a collection of notes and comments on the *Shi ji* by a Tang scholar Zhang Shoujie 張守節 (c. 600 AD).

> produce rain; feverishly they produce wind, randomly they produce fog; condensed they become hoar; dispersed they become dew; gathered they become clouds; standing up they become the rainbow; desolving they become a halo; separating they become a corona. The change of twenty four kinds of climates are all presided over by *Xuanyuan.*

In the *Jin shu*, *Xuanyuan* is directly mentioned as the spirit of Huangdi. Thus there is no doubt that the constellation *Xuanyuan* 軒轅 is the celestial presentation of Huangdi whose name is Xuan Yuan 軒轅. Even the shape of the constellation resembles the body of *huang long* (Yellow Dragon)!

This projection was in agreement with the way in which seasons were related to the Five Elements. The difficulty in the Five Element world view was how to correlate the four seasons with the five elements. One of the solutions was to take the last month of the summer *ji xia* to match the element of Earth[14]. In orientation, it would be south-west and that would be the position of *Xuanyuan* relative to the four palaces along the ecliptic (between *Taiwei* and *Wuche*).

Apparently the Central Palace is not for Huangdi, it is for the god even higher than Huangdi. During the Qin and Han, an imperial cult was instituted to create a new deity *Taiyi* 太一, the Supreme One. The Five *Di* were only the assistants of this supreme god[15]. The celestial residence of *Taiyi* was *Beiji*, the North Pole Office, just as the *Tianguan shu* said.

Thus we may conclude as follows. The Five Palaces were related to the movement of the sun along the ecliptic and mythologically with the Solar God Huangdi who was responsible for the change of the seasons. The four palaces along the ecliptic were the most natural way to divide the sky, to be directly related with four seasons and be the residences of the four *Di*. The establishment of the Central Palace reflected the evolution of philosophical and mythological ideas during the Han toward the completeness of the Five Element cosmology. *Taiyi*, the Supreme One was the personification of central power. This reflected the ideology of the Qin and Han time, of which the unification and central government were the main themes.

[14]See the *Huangdi neijing suwen* 黄帝内經素問 (Pure Questions on Medicine by Huangdi).

[15]See the *Shi ji*, chapter 28, *fengchan shu*, Monograph on the *feng* 封 and *chan* 禪 sacrifices.

6.3 *Five celestial courts*

According to the general astrological descriptions, five celestial courts can be recognized in the sky. They are the courts of *Taiwei* 太微, *Dajiao* 大角, *Tianshi* 天市, *Xin* 心 *xiu* and *Ziwei* 紫微[16]. Of course they were related to the Five Palaces that we have just discussed; the court of *Taiwei* corresponded with the southern Palace, the court of *Xin xiu* with the eastern Palace and the court of *Ziwei* with the central Palace. For the western and northern Palaces there were no courts. Two other courts were constructed, namely, the court of *Dajiao* and the court of *Tianshi*. Our interest here is to see how these courts were established by the three schools in constructing and naming constellations and to explore the cultural background.

6.3.1 Court of *Taiwei* (Privy Council)

In the *Tianguan shu*, the region of *Taiwei* was already described as a celestial court:

> *Taiwei* is the court of three luminaries; the twelve stars[17] which form the boundary of the court are vassals: on the west side they are generals, on the east side they are ministers; the four stars on the south side are law executors. In the middle front is the main gate, on both sides lateral gates. The six stars inside the gate are prince lords, still further inside are the five stars of *Wudi zuo* (Seat of Five *Di*); to the back, fifteen brilliant stars are gathering which are called *Langwei* (court officers); one big star beside them is the position of *Langjiang* 郎將, the general.

[16]Lü Zifang 呂子方 [1984] seems to be the first modern scholar to notice the existence of these celestial courts. He believes that these courts were created by different regional cultures in ancient China. That looked reasonable because people observed different stars to determine seasons and thus worshipped different stars. For instance, the Song tribe were supposed to descend from the Yin, thus they worshipped the Fire star *Huo* (α Scorpio), while the Jin tribe were supposed to descend from the Xia, and thus they worshipped the *Shen* stars (Orion). See *Lü Zifang kexueshi lunwenji* 呂子方科學史論文集 (Collection of Lü Zifang's Papers on History of Science; Sichuan People's Press, 1984, Vol. 2), pp.175-215.

[17]The *Shi Shi* counted 10 stars, which seems to be more reasonable according to the sky.

Shi Shi has almost the same description of this region of the sky. It has *Taiwei* 太微, *Huangdi zuo* 帝座, *Sidi zuo* 四帝座, *Wu zhuhou* 五諸侯, *Langwei* 郎位, *Langjiang* 郎將, etc., all recognizable in the *Tianguan shu*. Thus the court of *Taiwei* must have existed long before and must have been related with the southern Palace.

What kind of court was *Taiwei?* Alternative definitions from the above quotation are:

— Court of the three luminaries, the sun, the moon and the planets;

— Court of Huangdi or the Five *Di*;

— Court of *Taiyi*, the Supreme One.

The first definition is an astronomical observation. The second definition is due to the fact that Huangdi (or the Five *Di*) was the Solar God. However, it should also be the court of *Taiyi* because *Taiyi* had become the highest god to whom the Five *Di* were assistants.

After *Shi Shi* came *Gan Shi* which furnished this court even more like an imperial court. Mentioning the constellations which they constructed and providing a little more explanation will suffice to show the general ideas:

— *Sangong neizuo* 三公内座, the Inner Seats of the Three Excellencies;

— *Jiuqing neizuo* 九卿内座, the Inner Seats of the Nine Ministers;

— *Neiwu zhuhou* 内五諸侯, the Inner Five Prince-Lords[18];

— *Taizi* 太子, the Crown Prince;

— *Xinchen* 幸臣, the Favourite Court Servant, usually a eunuch;

— *Congguan* 從官, the Page;

— *Yezhe* 謁者, the Usher of the court;

— *Mingtang* 明堂, the Bright Hall;

— *Lingtai* 靈臺, the Spiritual Altar, the astronomical observatory.

One should notice the officialdom of the Zhou dynasty in this celestial court. It was not the actual Zhou officialdom but an idealized "Zhou officialdom" constructed by Han-time thinkers. The orthodox "neo-Confucianism" after Dong Zhongshu always advocated following the idealized example of the Zhou system of social order[19]. The real Zhou officialdom was not exactly known, which gave room for imagination – imagination based on

[18] The positions of state governors were usually taken by princes of the imperial family.

[19] Actually Confucius himself was an admirer of the Zhou social order. He said in his *Lun yu* 論語 (Confucian Analects): "Zhou had the advantage of surveying the two preceding dynasties. How replete was its culture! I follow Zhou."

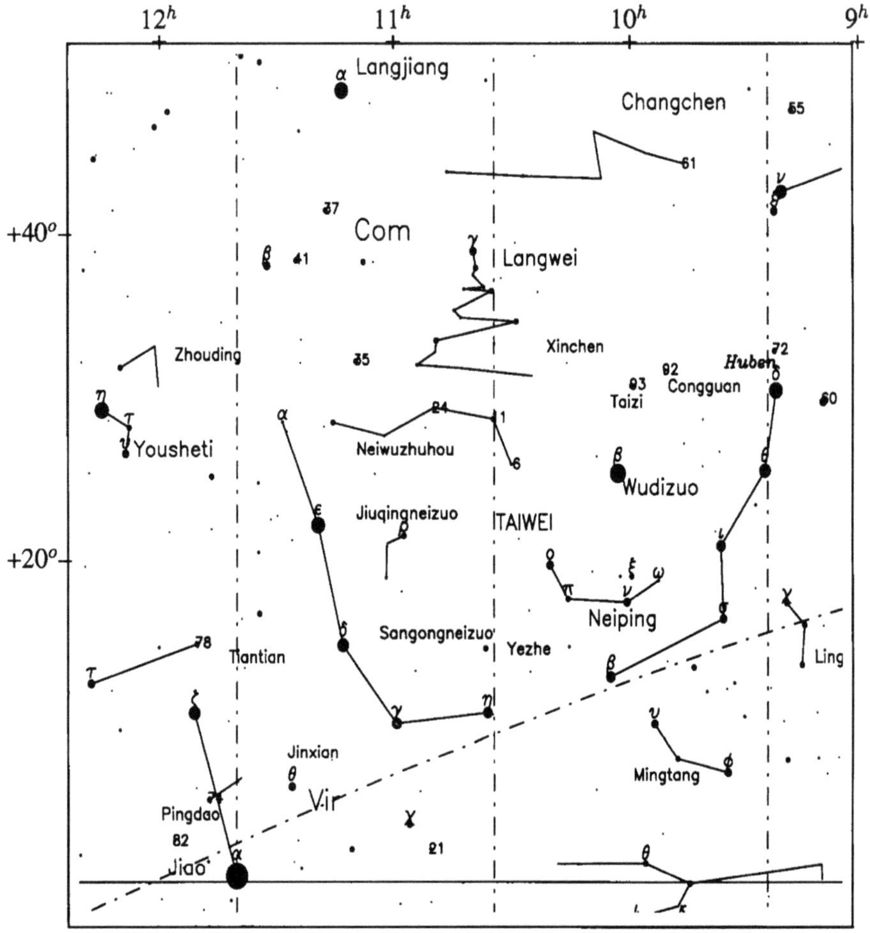

Figure 6.5: The Court of *Taiwei* 太微. in the sky. On the lunar calendar it corresponded to the early spring, about the middle of Febuary, when the Emperor announced the agenda for the coming year.

some Zhou traditions. The Three Excellencies and Nine Ministers system of "Zhou officialdom" had never been put into practice during the Zhou, it was not even mentioned in the *Zhou li* 周禮, (Rituals of the Zhou) which described the Zhou officialdom[20]. However, it was put in practice by Emperor Wu 武 (140 - 87 BC). According to practice, the Three Excellencies *san gong* 三公 were the *si tu* 司徒 (Minister of Labour), the *tai wei* 太尉 (Head of the Army) and the *si kong* 司空 (Minister of Constructions). This was apparently a Han invention[21]. At the end of the Former Han, Wang Mang 王莽 was enthusiastic with the "Zhou" order, because he wanted to be a "modern Zhou Gong 周公" by following the highly eulogized example of Zhou Gong assisting King Cheng 成 in establishing the Zhou[22]. Wang Mang was successful but not as loyal as Zhou Gong because he eventually took the Liu 劉 household (the Han dynasty) for his own (the Xin 新 dynasty)[23]. That was why the "Zhou officialdom" was projected in the sky by the *Gan Shi*. The ideological and political background were actually from Han times.

The construction of *Mingtang* and *Lingtai* during the reign of the Emperor Wu was also said to follow the Zhou tradition. The structure of the *Mingtang* and the ritual activities which should happen in it were typical Neo-Confucian symbolism. The sophisticated *Yin-yang* cosmology was incorporated into the *Mingtang* myths. The *Mingtang* was a miniature world in which the Order and Harmony of the universe were illustrated. Thus according to the *Yue ling* (Monthly Ordinances), in the first month of the year the new year calendar should be announced with solemnity by the emperor. This was indeed practiced during the Han. *Lingtai* was the Royal Observatory where the Son of Heaven (Emperor) could make contact, spiritually with his Ancestors.

---

[20]See Xu Fuguan 徐復觀, *Lianghan shixiang shi* 兩漢思想史 (History of Ideology during the Han; Hong Kong, 1975, vol II.)

[21]See the *Tong dian* 通典 (Comprehensive Institutes), a reservoir of source materials on political and social history by Du You 杜佑 from about 812 AD.

[22]Zhou Gong 周公 (Duke of Zhou) was the prime minister of King Wu 武 (1027 - 1025 BC). When King Wu was dying, he asked Zhou Gong to assist his son King Cheng to establish the Zhou dynasty. Zhou Gong became the regent of the young King and performed his duty faithfully.

[23]See the *Han shu*, ch.99, the Biography of Wang Mang.

According to the position of the anti-sun on the lunar calendar, *Taiwei* corresponded with the early spring during the Han. The celestial position of *Mingtang* corresponded with approximately the 10th of February which was the average date for the announcement of the new year's calendar in the Bright Hall *Mingtang*.

6.3.2 Court of *Dajiao*

Dajiao (α Bootis) is one of the brightest stars in the sky. In astrological texts, the constellations around *Dajiao* have always been described as a court of the Celestial King (*tian wang* 天王). We quote a few of these descriptions from the following books:

> from *Tianguan shu*: "*Dajiao* is the imperial court of the Celestial King. On both sides of it, escorting it like triples, are three stars of *Sheti* 攝提. *Sheti* is just in line with the handle of *(Bei) Dou*, which indicates the beginning of the seasons, thus it is also called *she ti ge* 攝提格 (*Sheti* the starting point)."
>
> from *Jin shu tianwen zhi*: "*Dajiao* is the seat of the Celestial King; it is also the pillar of the heavens, from which we count the degrees[24].
>
> from *Shi wei* 詩緯 [25]: "*Dajiao* is also called the mattress of Di, the ruling principles are formed there."[26]

It is obvious that *Dajiao* was to become the court of Heaven not only because of its brightness but also because of its function as the indicator of the beginning of the year. "To supervise celestial phenomena and to issue the calendar (*guan xiang shou shi* 觀象授時)" was the symbol of sovereignty, to which the emperor had to pay serious attention. The changing of the calendar would symbolize the changing of a dynasty. The full moon appeared at *xiu Jiao* in the early spring; *Dajiao* was nearby. *Dajiao* was also in line with the handle of *Beidou*, which had long been recognized as the indicator

[24]The original words are *zhen jing ji* 正經紀. Our translation is astronomical. It can be also translated as "to pronounce the principles to be followed". Both are actually based on the cosmological idea that man should follow the way of heaven.

[25]A "woof" book of the *Shi jing* (Book of Odes).

[26]See the *Kaiyuan zhanjing*, ch.65.

of the seasons. Therefore the area about *Dajiao* was described as the court of the celestial king[27].

All *Shi Shi*'s constellations along the line from the handle of *Beidou* to *Dajiao* had names of weapons, *Tianqiang* 天槍, *Xuange* 玄戈, *Zhaoyao* 招摇, and *Genghe* 梗河; they were apparently meant to safeguard the court. *Gan Shi*'s addition of *Dixi* 帝席 (Mattress of the Emperor) and *Zhouding* 周鼎 (Tripod of the Zhou) made the court even more complete. A separate group of seven Dukes, the *Qigong* 七公 might have belonged to this court.

6.3.3 Court of *Xin xiu*

Actually this court represents the eastern Palace. Not only *Xin xiu*, but also other *xiu* in this Palace were described as related to a celestial court.

About *xiu Jiao*, the *Jin shu tianwen zhi* says: "The two stars of *Jiao* form the celestial entrance, between which is the gate to the heavens, within (*nei* 内)[28] (the gate) is (situated) the celestial court. Thus through this gate passes the ecliptic, along which move the seven Luminaries."[29] Indeed, *Gan Shi* named two constellations here: *Tianmen* 天門 (Celestial Gate) and *Pingdao* 平道 (Flat Road) which led to the court. Moreover, *Gan Shi* indicated another asterism *Jinxian* 進賢 which meant prominent people approaching the court.

About other *xiu*, the *Jin shu* says:

> The four stars of *Kang* represent the inner court of the Emperor, looking after reports submitted from all over the country, setting disputes in the court, ...
>
> The four stars of *Di* represent the mansions of the Emperor and the rooms of the Empress and imperial concubines, i.e. the living apartment. ...

---

[27]It is interesting to note that in the Sumerian or ancient Babylonian Gilgamesh epic (c. 2000 BC) the solar god *Shamash* was at the position of the Chinese *Dajiao*. See Heidel [1963], pp.56-62, and Papke [1989], pp. 345-350.

From the Hindu-Vedic lunar lodge system (c.1000 BC) we know that the court of *Indra* corresponded with Chinese *Dajiao*. See Burgess [1935].

[28]*Nei* 内 literally means "within". But in the sky it actually means "to the east of" because the sun, moon, and planets pass the two stars of *Jiao* when coming from the west.

[29]Translation taken from Ho Peng Yoke; *Astronomical Chapter of Jin Shu*, Mouton and Co. Paris-The Hague, 1966, p.94.

> The four stars of *Fang* form the Bright Hall *ming tang*, where the Emperor sits and rules. ...
>
> The three stars of *Xin* form the throne of the King of Heaven. The central star (the Fire Star *Huo*), (also) known as *ming tang* (Bright Hall), (represents) the throne of the Son of Heaven. ...
>
> The nine stars of *Wei* form the inner palace and (represent) the mansions of the Empress and the imperial concubines. ...
>
> The four stars of *Ji* also denote the inner palace, the home of the Empress and the Imperial Concubines. ...[30]

Although duplication and confusion are visible in these astrological descriptions, the seven eastern *xiu* no doubt represented a celestial court.

The fact that the Fire star *Huo* was the "central star" of *Xin xiu* made *Xin xiu* a celestial court. At the time of Yao (about 2300 BC), this star was one of the four cardinal points. The Fire star became the time indicator during the Yin[31]. When the Fire star appeared above the eastern horizon at sunset, a new agrarian year began. Announcing the calendar was the privilege of the Emperor. Thus, astrologically, a celestial court was connected to *Xin xiu.*

6.3.4 Court of *Tianshi* (Celestial Market)

The court of *Tianshi* 天市 was apparently constructed after the appearance of the *Tianguan shu.* In *Tianguan shu*, *Tianshi* was described as only a small constellation of four stars: "To the north-east of *Fang xiu* and *Xin xiu* twelve stars are curving like a *Qi* 旗 (Flag). Inside the *Qi* are four stars of the *Tianshi.*" But a big wall-system was constructed here by Shi Shen's school, called *Tianshi Yuan* 天市垣. It thus resembled the court of *Taiwei* which also was surrounded by a wall. The twenty two stars on the wall were given names of provinces of the Han. They are:

— on the western wall: Song 宋, Wei 衛, Yan 燕, Donghai 東海, Xu 徐, Taishan 泰山, Qi 齊, Hezhong 河中, Jiuhe 九河, Zhao 趙, Wei 魏, Zhongshan 中山 and Hejian 河間;

---

[30] Translation taken from Ho Peng Yoke; *ibid*, p.95-98.

[31] See Pang Pu, "Huoli gouchen" 火曆鉤陳 (On the "Fire" calendar), *Zhongguo wenhua* 中國文化 (Chinese Culture), 1 (1990).

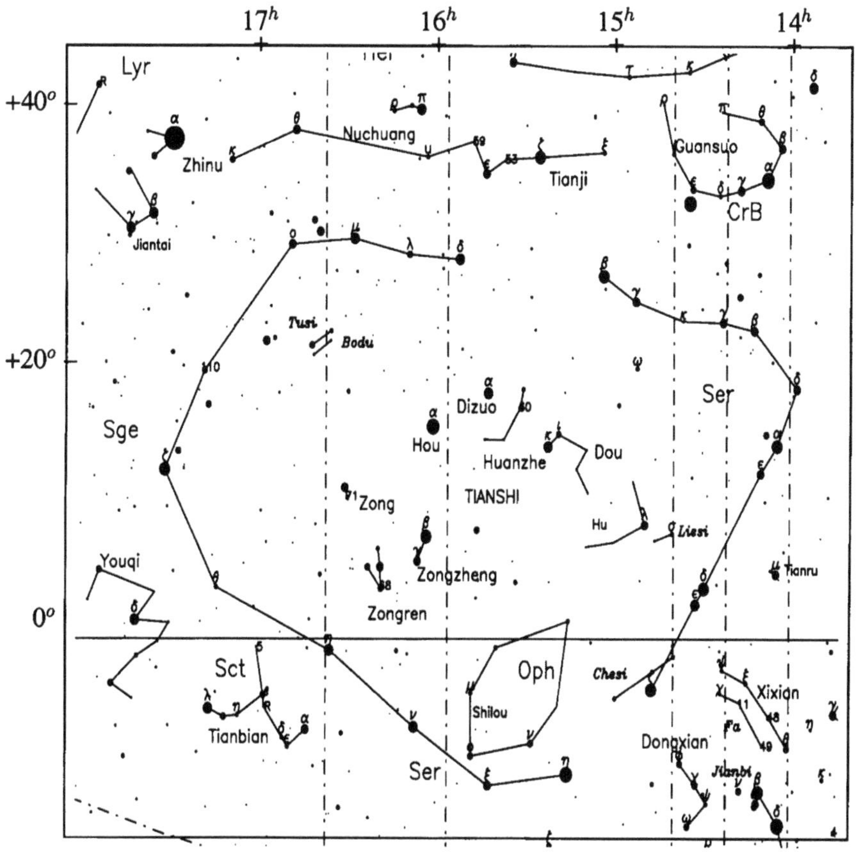

Figure 6.6: The *Tianshi Yuan* 天市垣, the Celestial Market. On the solar calendar it corresponded to the middle of November, the time of the autumn market.

— on the eastern wall: Han 韓, Chu 楚, Liang 梁, Ba 巴, Shu 蜀, Qin 秦, Zhou 周, Zheng 鄭 and Jin 晉.

These names help to determine the date of the construction of *Tianshi Yuan*. The coexistence of Jin 晉, Wei 魏, Zhao 趙 and Han 韓 indicates that this was after the time when the state Jin had been divided into the small states Wei, Zhao and Han in 481 BC[32]. The province of Hejian was only assigned as late as the end of the Former Han[33], which proves that *Tianshi Yuan* was constructed about the end of the Former Han.

The same idea of the *fen ye* 分野 (Field Division) was reflected in the construction of this court. All the stars on the wall were assigned to the regional states. The power of the Emperor was symbolized by the throne *Dizuo* 帝座 in the middle of this celestial court. He had his assistants watching over the country; thus the *Hou* 候 star was named to indicate such a function. The *Kaiyuan zhanjing* said: "*Hou* watches over the balance of *yin* and *yang* so that the country is stable."

This celestial "court" was somewhat different from others. It was essentially the Celestial Market. It was a typical autumn market (on the Solar Calendar) where the Imperial power was manifested. The tax system introduced by the Qin emperor was based on the introduction of standard measures. The celestial registration office *Tianji* 天紀 was actually the place to pay taxes. The ruling of the emperor was symbolized by standard measures; *Hou* 候 must also take care of them. The *Dou* 斗 and *Hu* 斛 constellations within *Tianshi* were indications of these measures.

Here *Shi Shi* had constellations such as *Dizuo* 帝座, *Hou* 候, *Huanzhe* 宦者, *Zongzheng* 宗正, *Zongren* 宗人 and *Zong* 宗 all relevant to a court. They represented the imperial court on the market.

Wuxian Shi added commercial aspects to the Market *Tianshi* 天市. They created the constellations *Liesi* 列肆, *Chesi* 車肆, *Bodu* 帛度 and *Tusi* 屠肆 to indicate shops. For more details see the appendix III.

[32]*Shi ji*, ch.39, *jin shijia* 晉世家 (History of the Jin State).

[33]Tan Qixiang 譚其驤, *Zhongguo lishi tidu* 中國歷史地圖 (Historical Atlas of China; Beijing, 1982.).

6.3.5 Court of *Ziwei Yuan*

The celestial north pole had a unique importance in ancient Chinese cosmology. Around it the Central Palace or Central Celestial Court was constructed. After the establishment of the central power during the Qin and the Han, imperial ideology was in favour of ideas which emphasized the central imperial authority. Thus the Central Palace was developed into an even more sophisticated celestial court.

The central court was higher in rank than other celestial courts and controlled the whole sky. This had already been well established with the constellations described in the *Tianguan shu* and those of Shi Shen's school. The North Pole Office (*Beiji*) had five stars; the brightest was the *Di* 帝 or Emperor, the others were called Queen, prince, crown prince and the celestial pivot. The last one apparently indicated where the real celestial pole was supposed to be. Thus this central court symbolized the Imperial Court.

At the backdoor of the court was a Carriage Road (*Gedao* 閣道) passing over the Celestial River to the ancestor shrine of *Yinshi* 營室 (*Shi* 室 *xiu*) where the Emperor's summer palaces were located. Moreover the central court used *Beidou* as the "wagon of the Emperor" which revolved around the pole and thus controlled the whole sky.

To perfect the functioning of this Center of Imperial Power, Gan De's school constructed more constellations which represented officials, institutions, facilities and deities in connection with the court. A list of these constellations with brief explanations will show the total picture[34]:

— *Tianhuang dadi* 天皇大帝, High God of Heaven. This name combined four Chinese characters which respectively meant Heaven, Emperor, Greatness, and Mighty Great Lord. A combination of these holy words was supposed to have an even stronger indication of Supremity.

— *Sifu* 四輔, Four Advisors of the Emperor, according to the idealized Zhou officialdom.

— *Yinde* 陰德, Virtue of the *yin* 陰 force.

— *Liujia* 六甲, Six *jia* (the first Stem of the Chinese ten Stems of counting). It became deified because a type of numerological art of prognostication, called *qi men dun jia* 奇門遁甲, appeared during the Han (see Appendix II, 5).

[34]For more details see Appendix II.

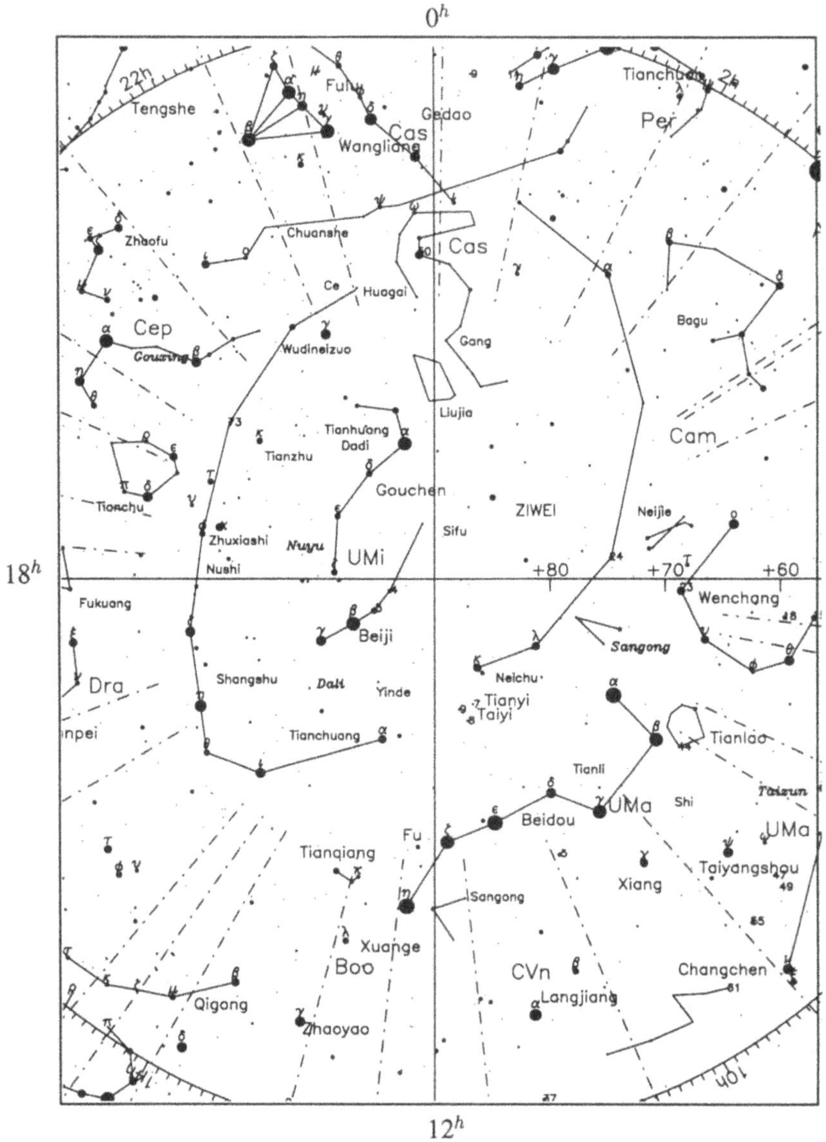

Figure 6.7: The *Ziwei Yuan* 紫微垣, the central celestial court.

— *Shangshu* 尚書, Secretary to the Emperor.

— *Tianzhu* 天柱, Pillars for Royal Announcements.

— *Zhuxiashi* 柱下史, Officer in charge of public announcements.

— *Huagai* 華蓋, Canopy of the Emperor.

— *Wudi neizuo* 五帝内座, Inner Seats of Five *Di* 帝.

All these constellations can hardly be identified in the polar region, where not many visible stars exist. They were constructed from imagination so as to establish the central court. And in this picture we again see traces of Han-time philosophies and ideas.

6.4 *Other significant groups of constellations*

6.4.1 Sun star and Moon star

Gan De's school has two asterisms called Sun Star and Moon Star, each consisting of only one star. They are not bright, but their designation has an important meaning.

The Sun Star *Ri* 日 and the Moon Star *Yue* 月 were located near the ecliptic and opposite to each other in the sky. This indicates that *Gan Shi* knew that the sun and the full moon were in opposition. Thus these two stars were named deliberately by astrologers to indicate the relation between the sun and the full moon. This insight was not general accepted until the appearance of the *hun tian* cosmological theory at the late Former Han. Therefore the registration of these two stars was not earlier than this[35].

Probably more significant were the exact positions of the Sun star (4 Lib) and the Moon star (37 Tau). They divided the year into two *equal numbers of days,* although the differences in longitude of the two stars measured along the ecliptic were quite unequal.

---

[35]Zhang Heng (79-138) explained in the *Ling xian* 靈憲 that the lunar eclipse was caused by the earth's shadow *an xu* 闇虛. This indicates that the relative position of the sun and the moon with respect to the earth was clearly understood at his time. He was the one who also wrote the *Huntianyi zhu* 渾天儀注 (Notes on the Armillary Sphere) which described the *hun tian* theory in general.

	Difference in Longitude	Number of days	Season
Yue to *Ri*	175 degrees	182 – 183	Summer-half
Ri to *Yue*	185 degrees	182 – 183	Winter-half

Apparently the Han astronomers already noticed the unevenness of the solar motion; during the "summer-half" the sun moves slower along the ecliptic than during the "winter-half"[36]. Although they did not understand its mechanical origin[37], this fact is a demonstration of their observational skill and accuracy.

Secondly, the mythology of sun and moon was considered as well. In mythology, the sun is accompanied by a three-legged crow and the moon by a toad. Gan De's school commented on these two stars like this: "The sun pronounces the order of Virtue, its luminosity produces a flying crow; the moon illustrates the measures of Punishment, its luminance produces a toad."[38] The Sun Star *Ri* was near *xiu Fang* which belonged to the Eastern Palace; the Moon Star *Yue* was near *xiu Mao* which belonged to the Western Palace. This was again in agreement with mythological descriptions about the position of the sun and moon; the sun was always related to the east and the moon to the west.

It is also interesting to note that *xiu Fang* and *xiu Mao* were exactly at the autumn and spring equinoxes at the time of Yao Di, about 2300 BC. *xiu Fang* and *xiu Mao* were among the four cardinal asterisms recorded in the Canon of Yao. This again refers to the Yao time astronomical tradition. It was known to some astrologers of the Han time. It is possible that Han astrologers absorbed this ancient tradition into their new system of constellations. The Sun Star and the Moon Star are typical examples. The names of

[36]Historically the unevenness of the sun's motion was not recognized in China until the time of Zhang Zixin 張子信 (in the Northern Qi 齊 dynasty, about 570 AD.), but the fact that the equinoxes are not placed at exactly equal intervals between the solstices may have been implicit in data available to Han astronomers. See Needham, *Science and Civilization in China*, vol. 3, section 20, pp.393-394.

[37]The explanation came 1600 years later from Johannes Kepler. It is due to the eccentricity of the earth-orbit around the sun. The Han-time ratio of 185 to 175 corresponded with an eccentricity of 0.014 whereas the modern value is 0.0167. The center of the winter-half of the year is reached on the 2nd of January, when the earth reaches its perihelium.

[38]The *Kaiyuan zhanjing*, ch. 69.

some constellations, if related to this date of 2300 BC, are more reasonable according to the seasonal projection scheme[39].

6.4.2 Autumn harvest picture

A harvest picture was visible in the southern part of sky where *Lou* 婁 *xiu* and *Wei* 胃 *xiu* were located. During the Han the spring equinox was at *Lou*. According to the Lunar calendar, this part of the sky was seasonally connected with September and October, which was the season for harvesting rice. The harvest picture was made visible by *Shi Shi*'s constellations *Tiancang* 天倉 (Celestial Barn), *Tianjun* 天囷 (Celestial Grain Stores), and *Tianlin* 天廪 (Celestial Granary). The harvest picture dominated in *Wei xiu*. Sima Qian said in *Tianguan shu* "*Wei* is the celestial barn." *Shi Shi*'s constellations were more specific.

Gan Shi continued to fill this region with various objects of country life, like a sewer for manure *Tianhun* 天溷 and a screen to keep it out of sight *Waiping* 外屏. We also see a reaping sickle *Fuzhi* 鈇鑕 and some ricks of grain (with straw) *Tianyu* 天庾 in the field.

6.4.3 Hunting picture

Beyond *xiu Wei* and *Lou* come the sections of *xiu Bi*, *Zui* and *Shen*. This part of the sky had been noticed since very early times; *Shen* was one of the most ancient constellations. These three *xiu* were all connected with hunting. *Shen* symbolized a tiger, *Zui* a bird, and *Bi* a rabbit-trap. On the star painting of the Han tomb that picture was clearly shown. See Figure 6.1. Also the *Kaiyuan zhanjing* quoted Xi Meng saying: "if hunting is happening, one should investigate *xiu Bi*." Thus this part of the sky was definitely related to hunting activities. According to the Lunar Calendar, between the time of Yao and the Han, this part of the sky was connected with the months of November and December or the beginning of winter, the best season for hunting. Shi Shen's constellations in this region like *Lang* 狼 (Wolf), *Hu* 弧 (Bow) and *Yeji* 野雞 (Pheasant) were also obviously related to the hunting picture.

---

[39] A shift of one month results from the difference between this date and the Han time.

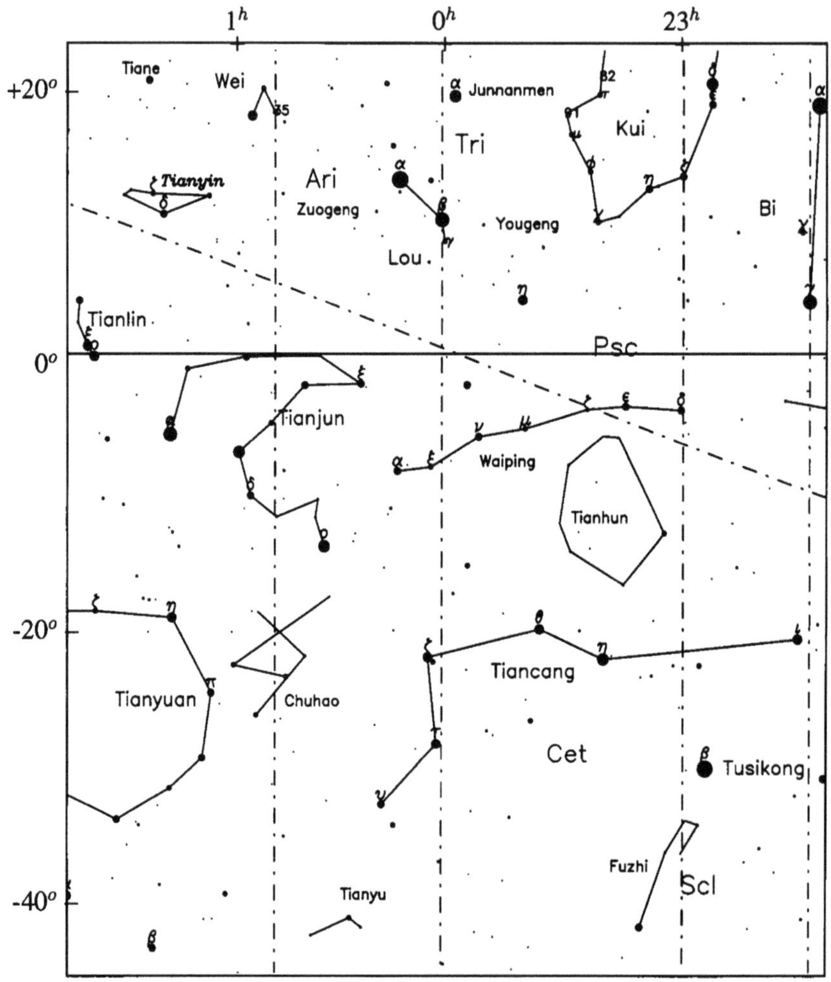

Figure 6.8: The part of the sky related to the autumn harvest.

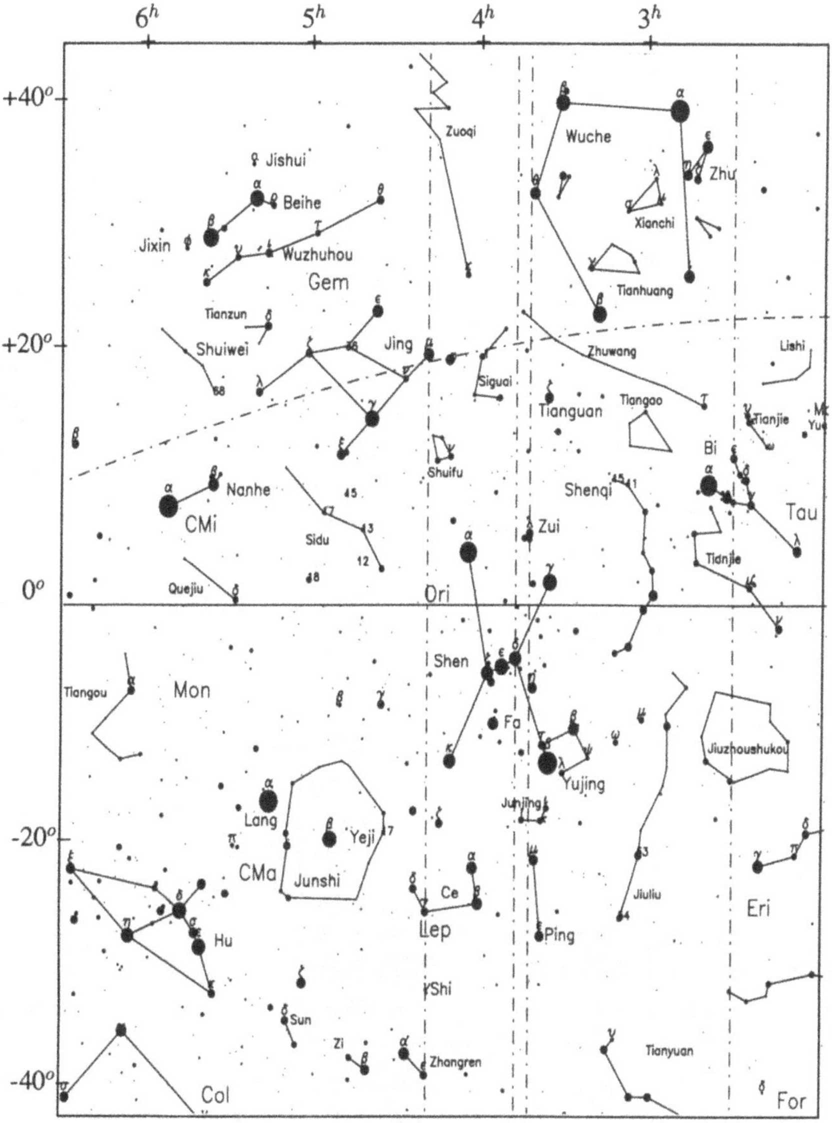

Figure 6.9: The part of the sky related to the hunting.

Ancient hunting traditions were followed by the Han during the imperial hunting ceremony. The ceremony, which was called *tian lie jiang wu* 田獵講武 (Field Hunting and Military Training), usually happened during the middle of the winter (December). The ceremony had two purposes. One was to bring an offering to gods. The other was to inspect the army, since thousands of soldiers were needed to enclose an area for the hunting party.

To understand this part of the sky, let us first introduce a description of this ceremony in the *Xin tang shu liyi zhi* 新唐書・禮儀志 (Treatise on Rituals and Ceremonies of the New History of the Tang). We summarize the event as follows:

> The Field Hunting by the Emperor happened in the middle of the winter. Ten days before the game-preserve was enclosed. Three days before the hunt the banners were erected in the fields. The day before the event, at twilight, generals and soldiers were put in position. Then they started to rouse the game in the field from two sides, and the Emperor rode into the field, all generals following. After driving the animals three times the Emperor shot the game first, after which the lords and dukes followed; at the end the peasants joined in. Big animals were reserved for the whole community, smaller ones were granted to individuals. The best animals were sacrificed to the ancestors, the lesser ones were served to the guests and the poorest ones were sent to the kitchen.

In the context of such a description, most constellations in this part of the sky become understandable. *Zuoqi* 左旗, *Tianjie* 天節, *Shenqi* 参旗 and *Jiuliu* 九游 were all sorts of banners to indicate nobilities and ranks, *Junshi* 軍市 was the market for soldiers. *Yujing* 玉井 was the well for the nobles, *Junjing* 軍井 was the well for the soldiers. There were so many guests that the *Jiuzhou shukou* 九州殊口 (interpreters) were needed. Peasantry was represented by a family of three generations: *Zhangren* 丈人 (husband man), *Zi* 子 (son) and *Sun* 孫 (grandson). *Siguai* 司怪 (the Animal Tamer) was in charge of the hunting dogs and knew from the reactions of his dogs where the game was. Both Shi Shen's and Gan De's schools contributed to make this hunting picture complete.

Figure 6.10: A hunting picture of the Emperor Qianlong 乾隆 of the Qing dynasty, painted by Giuseppe Castiglione who served at the Qing Court from 1715 to 1766. This picture illustrates the scene of the Field Hunting.

6.4.4 The Cavalry Camp

The sky to the south of the *xiu Jiao, Kang, Di* and *Fang* was a complete picture of a cavalry camp. *Tianguan shu* describes this part of the sky as a large cavalry arsenal *Ku* 庫 with a separate tower *Lou* 樓 at the southern entrance; many cavalry officers are mentioned.

Shi Shi described the constellations *Kulou* 庫樓 (Arsenal), *Qiquan* 騎官 (Cavalry Officer) and *Jizu* 積卒 (Group of Soldiers).

Gan Shi constructed *Zhewei* 折威, an officer in charge of executions; *Zhenche* 陣車, a battle chariot; *Qizhen jiangjun* 騎陣將軍, the commander of the cavalry and *Cheqi* 車騎, the crew for battle chariots.

Wuxian Shi introduced *Dunwan* 頓頑, the officer to maintain discipline in the army. Moreover they created *Congguan*, an officer to take care of health and medicine needed by the troops. An important contribution was *Tianfu* 天輻, a pile of reserve spokes to repair broken wheels of carriages.

During the Han, there were continuous incursions of the northern nomadic Xiongnü 匈奴 people into the valleys of the Yellow River. They usually came to plunder the harvest in autumn. That was a major problem for the agrarian Han. When Han Wu Di began his reign, he was determined to get rid of this trouble. At first a mounted army like that of the Xiongnü was established. Then the army went to war against the Xiongnü[40]. Between 130 and 119 BC there were annual attacks of Chinese cavalry led by the famous general Huo Qubing 霍去病. At last the Han Emperors succeeded in driving the Xiongnü to the far north-west[41].

The cavalry camp in the sky was apparently the projection of these military activities. According to the Solar Calendar it corresponded to the autumn, exactly the season when these expeditions happened[42].

[40]See the *Han shu*, ch.94, *Xiongnü zhuan* 匈奴傳 (Descriptions of the (ancient) Hun people).

[41]In 53 BC, the Khan of the western Xiongnü surrendered and in 36 BC the northern Khans were subdued. The western Xiongnü fled in the western direction and arrived in 450 AD in Hungary led by their Khan Attila. See A.F. Hulsewe; *Weltgeschichte*, Propyläen Verlag, 1960-65, p.560.

[42]According to the *Han shu*, most of these expeditions were organized in autumn.

6.5 *The structure of the Chinese sky re-examined*

Apart from structures and groups in the Chinese sky discussed above, there are two other features worthy of notice.

6.5.1 Inner and Outer Constellations

All constellations, except the twenty eight *xiu*, were distinguished as being either of the *zhong guan* 中官 (inner or central constellations) or of the *wai guan* 外官 (outer or peripheral constellations). The inner constellations were located in the central region to the north of the belt of the 28 *xiu*; while the outer constellations were located to the south of the belt. Because of the diurnal motion of the celestial sphere, the inner constellations rose high in the sky towards the zenith, while the outer-constellations stayed nearer the southern horizon. In Chinese cosmology, the region about the north pole was the noblest part of the sky; the closer to the north pole, the nobler the region.

Thus the inner constellations were considered to be nobler than the outer ones. This was obviously shown by their names:

— Inner constellations received names of nobility like emperors and kings or generals and officials. Also imperial institutions and facilities as well as emblems and jewelry could be expected among the inner constellations.

— Outer constellations received mainly names from life in the countryside and remote territories, or from the army.

Peasants had always the lowest social status in ancient China; thus when projected in the sky, peasant related objects could only be placed in the most southern parts of the sky. The army, which was primarily recruited from peasants, was stationed at remote boundary areas to defend the empire; thus they were also projected in the lowest part of the sky. Social classification was clearly reflected in the names of constellations.

6.5.2 About the sequential order of constellations

It was natural that the sequence of the 28 *xiu* followed the motion of the sun, moon and planets. In the *Kaiyuan zhanjing* the constellations of Shi Shen's

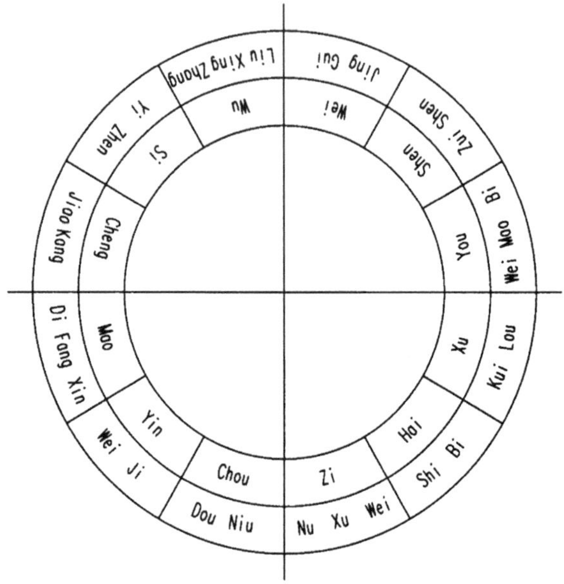

Figure 6.11: The correspondence between the Twelve Branch segments and the 28 *xiu*.

school followed the same sequence. They started from *Jiao xiu* and went on to *Zhen xiu*; then, through the court of *Taiwei*, via the Three Steps (*San Tai* 三臺), they went up to the central court.

But Gan De's school organized its constellations in a different way. According to the sequence presented in the *Kaiyuan zhanjing*, its inner constellations started from the polar area, then followed the Carriage Way (*Gedao* 閣道) from the central court *Ziwei* to *Shi* and *Bi* 壁 *xiu*, and then went on in the reverse direction along the ecliptic. Thus the central court was more emphasized by Gan De's school, which was certainly according to Han philosophy.

Moreover, the reverse sequence of the constellations of *Gan Shi* had more implications. It actually followed the sequence of another division system of the sky – the Twelve Branch *shi'er chen* 十二辰 Division. Here the sky was divided into twelve equal segments along the equator, but they were numbered in a sequential order opposite to the annual movement of the sun. They were named after the twelve Branches, in the sequence of *zi*

子, *chou* 丑, *yin* 寅, *mao* 卯, *chen* 辰, *si* 巳, *wu* 戊, *wei* 未, *shen* 申, *you* 酉, *xu* 戌 and *hai* 亥. This division apparently started from the sector where the winter solstice was, and proceeded in the opposite direction to the solar motion. The correspondence between the Twelve Branches and the 28 *xiu* is shown in Figure 6.11.

The Twelve Branch division of the sky was mainly for astrological purposes and was invented rather late, probably during the Han. It was extensively used on the Divination Board (*shi pan* 式槃), a form of prognostication which also developed during the Han. The fact that Gan De's constellations followed this division system strengthens our opinion that Gan De's school was more astrologically oriented and its constellations were constructed later than those of *Shi Shi*. The general picture of the sky of Gan De's school was a product of the Han period.

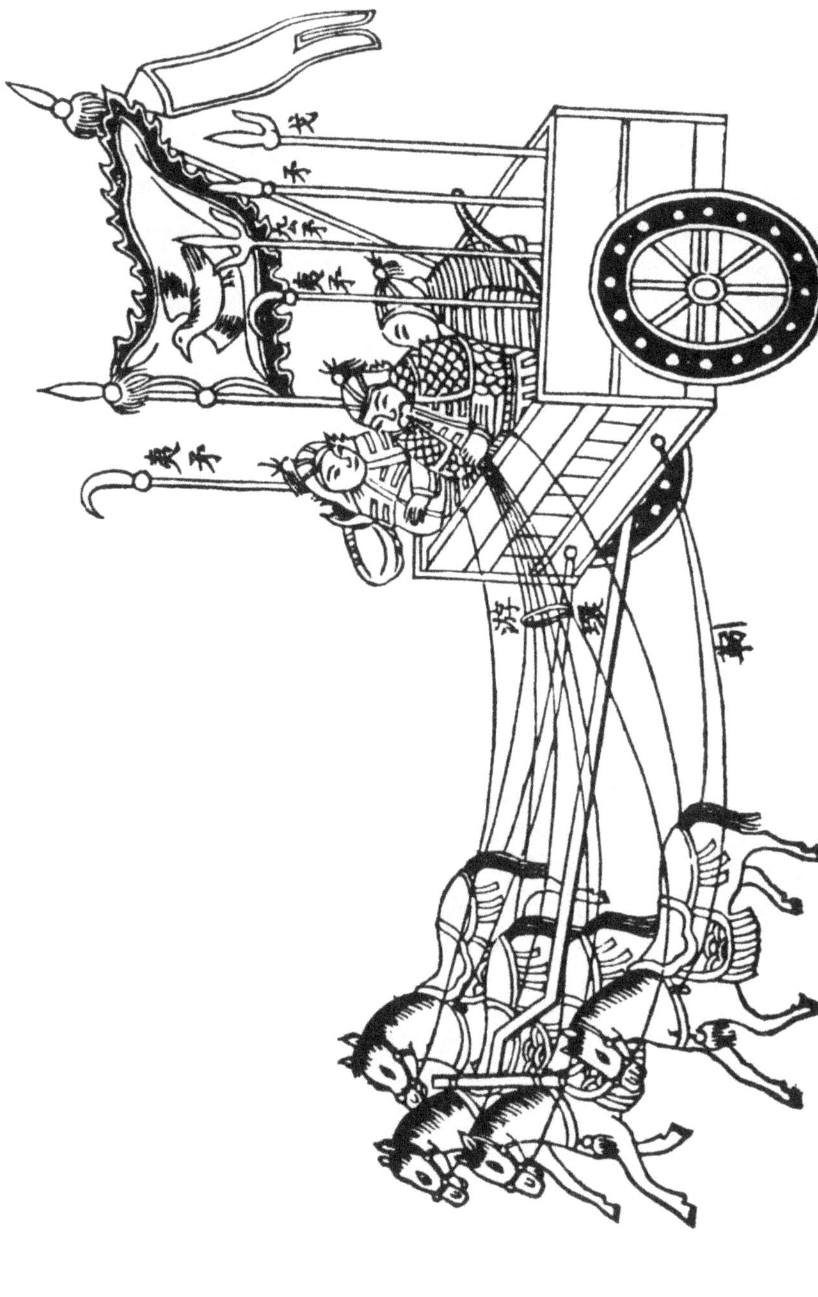

A picture of a battle chariot in ancient China. In the sky there are three constellations *Wangliang*, *Zhaofu* and *Xizong*, named after three legendary charioteers: Wang Liang, Zhao Fu and Xi Zong. There are also an arsenal *Chefu* and a camp of battle chariots near *Kulou* in Centaurus.

APPENDIX I

THE CONSTELLATIONS OF SHI SHI

In this appendix we present a brief historical approach to each constellation of *Shi Shi*. The constellations are listed in the same sequence as in the *Kaiyuan Zhanjing*. Each entry gives particulars in the following order:

1. The star name in *pinyin* spelling, together with Chinese characters.
2. The identification of the leading star of the constellation according to our study in Chapter 3.
3. The number of stars in the constellation according to the *Kaiyuan zhanjing*.
4. A brief translation or circumscription of the name of the constellation.

5. Explanations and comments.

By "(SS)", "(GS)" or "(WS)" after a star name we indicate the school to which this constellation belonged: *Shi Shi*, *Gan Shi* or *Wuxian Shi*.

1. *Sheti* 攝提; η (Boo), 6 stars;
   regulators; the right and left helpers of the Celestial King *Dajiao*; they also indicate the beginning of the seasons (Sima Qian: *Tianguan shu*, hereafter *TGS*).
2. *Dajiao* 大角; α (Boo), 1 star;
   the Great Horn of the Celestial King; the area around *Dajiao* is described as a Celestial Court (*TGS*).
3. *Genghe* 梗河; ρ (Boo), 3 stars;
   a shield, the weapon supposed to protect the Celestial Court of *Dajiao*. All the following weapons are supposed to have the same function.

4. *Zhaoyao* 招搖; γ (Boo), 1 star;
   a kind of weapon, sword or spear; it was mentioned in the bamboo book *Yunmeng qinjian rishu* 雲夢秦簡日書 from the 3rd century BC.
5. *Xuange* 玄戈; λ (Boo), 1 star;
   a halberd.
6. *Tianqiang* 天槍; κ (Boo), 3 stars;
   a celestial spear.
7. *Tianpei* 天棓; ι (Her), 5 stars;
   a celestial flail to thresh grain or another spear (*TGS*).
8. *Nüchuang* 女床; π (Her), 3 stars;
   a woman's bed. The *Jin shu* says that it represents the ladies of the inner palace, and governs the affairs of women.
9. *Qigong* 七公; β (Boo), 7 stars;
   seven dukes.
10. *Guansuo* 貫索; β (CrB), 9 stars;
    a chain, a loop of links; representing the gaols for common people (*TGS*). There is another prison for noble people, *Tianlao* 天牢, near *Beidou*.
11. *Tianji* 天紀; ξ (CrB), 9 stars;
    celestial discipline. *Jin shu* mentions an office for registration of market affairs.
12. *Zhinü* 織女; α (Lyr), 3 stars;
    the spinning damsel. According to Sima Qian she is a weaving princess of the celestial court.

13. *Tianshi* 天市; ζ (Oph), 22 stars;
    the celestial market.

    The 22 stars formed a wall-system in the sky. Names of the Han provinces (or states) were given to these stars by *Shi Shi*. On the west wall from South to North they were: Song 宋, Wei 衛, Yan 燕, Donghai 東海, Xu 徐, Taishan 泰山, Qi 齊, Hezhong 河中, Jiuhe 九河, Zhao 趙, Wei 魏, Zhongshan 中山 and Hejian 河間;

    On the east wall from South to North they were: Han 韓, Chu 楚, Liang 梁, Ba 巴, Shu 蜀, Qin 秦, Zhou 周, Zheng 鄭 and Jin 晉.

Figure 7.1: A stone-relief from the Eastern Han dynasty, showing the Spinning Damsel *Zhinü* 織女 (Vega) and the Sun. When the Sun approached the constellation *Zhinü*, it was mid-winter, the beginning of the year on the calendar of the Zhou.

Although most states existed during the Zhanguo period (475-221 BC), the presence of Hejian indicates that this construction originates from the Han since this provincial name did not exist before the Han.

14. *Dizuo* 帝座; α (Her), 1 star;
the seat of the emperor.
15. *Hou* 候; α (Oph), 1 star;
Hou literally means "watching over". It could represent a kind of assistant of the Emperor, for example an astrologer who "supervises the balance of the *yin* and the *yang* in order to keep the country stable." See the *KYZJ*, ch.68.
16. *Huanzhe* 宦者; 60 (Her), 4 stars;
several eunuch administrators.
17. *Dou* 斗; κ (Her), 5 stars;
a measure for grain or liquid.
18. *Zongzheng* 宗正; γ (Oph), 2 stars;
an official who was in charge of the royal clan.
19. *Zongren* 人; 68 (Oph), 4 stars;
the assistants of *Zong Zheng*.
20. *Zong* 宗; 71 (Oph), 2 stars;
an important ancestor; during the Han it could mean Liu Bang 劉邦, the first emperor of the Han.

21. *Dongxian* 東咸 and *Xixian* 西咸; ψ (Oph), 8 stars;
buildings enclosing a place for judging and punishing of market frauds. They are literally the eastern and western walls enclosing the place where "harmony" is restored.

22. *Tianjiang* 天江; θ (Oph), 4 stars;
the celestial river, synonymous with *tian he*, the Milky Way.

23. *Jianxing*建星; ξ (Sgr), 6 stars;
city with a banner according to *Jin shu.* Sima Qian mentions a flag.

24. *Tianbian* 天弁; 5 (Aql), 9 stars;
the celestial casque, symbolizing authority. *Jin shu* mentions regulators of the Celestial Market.

25. *Hegu* 河鼓 and *Youqi* 右旗; α (Aql), 12 stars;
a big drum at the river (3 stars) together with a banner (*Youqi*) on the right side(9 stars).

26. *Lizhu* 離珠; 1 (Aqr), 4 stars;
four bright pearls.

27. *Hugua* 匏瓜; ζ (Del), 5 stars;
a hard and dry melon.

28. *Tianjin* 天津; δ (Cyg), 9 stars;
a celestial ford, the shallow place in the celestial river. The Chinese regarded the Milky Way as the Celestial River *tian he* 天河, so the place with few stars was naturally taken as the shallow place in the river. This can be clearly seen in the sky.

29. *Tengshe* 螣蛇; α (Lac), 22 stars;
the snake of heaven or a flying serpent (*Jin shu*). According to the *Kaiyuan zhanjing* (ch.65) there are creatures living in the water, mating with river turtles (*Bie*) and tortoises (*Gui*).

30. *Wangliang* 王良; β (Cas), 5 stars;
Wang Liang, a legendary charioteer with a whip (*Ce* 策). When Wang Liang whips his horses the fields are filled with running chariots (*TGS*). He was a charioteer of the State of Jin. One can read about Wang Liang (and Zhao Fu) in the *Zuo zhuan* 左傳.

31. *Gedao* 閣道; ϕ (Cas), 6 stars;
a stepped road through mountainous territory (*TGS*).

32. *Fulu* 附路; υ (Cas), 1 star;
    an auxiliary road to *Gedao*; also represents the chief assistant maintaining roads.
33. *Tianda jiangjun* 天大將軍; γ (And), 11 stars;
    the celestial Great General. The *Sui shu* says that the bright star (γ And) represents the Great General of Heaven whereas the other 10 stars are his subordinates, ready to follow him.
34. *Daling* 大陵; 9 (Per), 8 stars;
    a mausoleum.
35. *Tianchuan* 天船; η (Per), 9 stars;
    a celestial boat in the celestial river.
36. *Juanshi* 卷舌; ν (Per), 6 stars;
    a curled tongue, probably representing a slanderous woman.
37. *Wuche* 五車 and *Sanzhu* 三柱; ι (Aur), 14 stars;
    five chariots with three sets of poles to bind horses. Sima Qian mentioned Huangdi and the Four *Di* in a camp, probably the hunting camp.
38. *Tiangguan* 天關; ζ (Tau), 1 star;
    a celestial pass along the ecliptic which opens and closes.
39. *Nan bei he* 南北河; β (CMi), 6 stars;
    a combination of *Nanhe* 南河 (Southern River) and *Beihe* 北河 (Northern River) in one constellation. Another form of the name of this constellation is *Nan bei hexu* 南北河戍 (Southern and Northern Valleys of the River) border area where the garrison troops produce food for themselves, introduced during the Han. See the *Jin shu*.
40. *Wuzhuhou* 五諸侯; θ (Gem), 5 stars;
    five Lords of Heaven.
41. *Jishui* 積水; o (Gem),1 star;
    a store of water to make wine.
42. *Jixin* 積薪; ϕ (Gem), 1 star;
    stored wood.
43. *Shuiwei* 水位; 68 (Gem), 4 stars;
    a water level marker.
44. *Xuanyuan* 軒轅; α (Leo), 17 stars;
    Jin shu says it is the spirit of Huangdi and the body of the Yellow Dragon (also representing Huangdi). It is also the solar wagon.

45. *Shaowei* 少微; 53 (Leo), 4 stars;
a group of eminent men of high rank, welcoming Huangdi on his way towards the Court of *Taiwei*.

46. *Taiwei* 太微; β (Vir), 10 stars;
Privy Council, the Spring Palace of Huangdi where he receives the four *Di* in Privy Council for consultation about the New Calendar. On the Lunar Calendar it corresponded with the early spring.

The ten stars of *Taiwei* formed a wall-system, *Yuan* 垣 surrounding the palace; thus *Taiwei* was also called *Taiwei Yuan*. Each star had its own name. On the west wall from South to North they are:
— *You zhifa* 右執法; right law administrator;
— *Xi shangjiang* 西上將; west upper general;
— *Xi cijiang* 西次將; west lower general;
— *Xi cixiang* 西次相; west lower minister;
— *Xi shangxiang* 西上相; west upper minister;

On the east wall from South to North they are:
— *Zuo zhifa* 左執法; left law administrator;
— *Dong shangxiang* 東上相; east upper minister;
— *Dong cixiang* 東次相; east lower minister;
— *Dong cijiang* 東次將; east lower general;
— *Dong shangjiang* 東上將; east upper general.

Shi Shi did not mention these names. It was a later construction by astrologists to illustrate the Court of *Taiwei*. In the *Jin shu* all these names were given.

47. *Huangdi zuo* 黄帝座; β (Leo), 1 star;
the seat of the Solar God Huangdi. The *Jin shu* mentions here *Wudi zuo* 五帝座, the seats of five *Di*. Actually it was a combination of *Huangdi zuo* and *Sidi zuo*. (see below)

48. *Sidi zuo* 四帝座; invisible, 4 stars;
the seats of the four *Di* of the four Celestial Palaces at the four Cardinal Directions. They surrounded the throne of Huangdi and were called the Green *Di* of the Eastern Palace, the Red *Di* of the Southern Palace, the White *Di* of the Western Palace and the Black *Di* of the Northern Palace.

49. *Neiping* 内屏; ξ (Vir), 4 stars;
the inner screen to protect Huangdi and the four *Di*.

50. *Langwei* 郎位; γ (Com), 15 stars;
Gentlemen serving in the Court, particularly the body-guards.

51. *Langjiang* 郎將; α (CVn), 1 star;
Captain of the body-guards.

52. *Changchen* 常陳; 61 (UMa), 7 stars;
the standing rearguards of the Court of *Taiwei.*

53. *Santai* 三臺; ι (UMa), 6 stars;
three steps between the Court of *Taiwei* and the Court of Ziwei. *Santai* showed three levels of education, beginning with the lowest step *Xiatai* 下臺, followed by a step for middle class dignitaries *Zhongtai* 中臺 and finally by a step for high dignitaries *Shangtai* 上臺 (*Jin shu*)

54. *Xiang* 相; χ (UMa), 1 star;
the prime minister (executive function) acting outside the Celestial Court *Ziwei.*

55. *Taiyangshou* 太陽守; ψ (UMa), 1 star;
the General in charge of the *yang* valve, symbolising Chief Minister. (*Jin shu*)

56. *Tianlao* 天牢; 44 (UMa), 6 stars;
the prison for noblemen.

57. *Wenchang* 文昌; *o* (UMa), 6 stars;
the administrative center of the Celestial Government. Sima Qian mentioned six functionaries, represented by six stars. Afterwards *Jin shu* specified their functions.

58. *Beidou* 北斗 and *Fu* 輔; α (UMa), 8 stars;
the Northern Dipper to indicate the time and regulate the seasons. Sima Qian described it as follows: "It serves as the chariot of the emperor *Di* and controls the four cardinal points of the world by revolving around the centre; it divides the *yin* and the *yang* and regulates the four seasons; it maintains equilibrium between the five phases; it regulates the moving of the celestial objects; it decides about the epochs of all periodic evolutions (*ding zhu ji* 定諸紀) and the calendar". Sima Qian's report reflected the Confucian philosophical thinking of *Shi Shi*. *Fu* 輔 was the assistant of the Emperor to operate the valve regulating the *yang* force (*Kaiyang* 開陽).

59. *Ziwei* 紫微; κ (Dra), 15 stars;
the Celestial Purple Palace enclosed and surrounded by a wall indicated by 15 stars, *Ziwei Yuan* 紫微垣. The 15 stars were named in the same fashion as on the wall of *Taiwei*. They were purely from the astrologer's imagination.

60. *Beiji* 北极, *Gouchen* 鉤陳; β (UMi), 5 stars;
the North Pole Office consisting of five stars and centered by *Di*, the Emperor.

61. *Tianyi* 天一; 7 (Dra), 1 star;
Heavenly Unity.

62. *Taiyi*一; 8 (Dra), 1 star;
August Unity; Supreme One.

63. *Kulou* 庫樓; δ (Cru), 29 stars;
a large cavalry arsenal *Ku* 庫 with a separate tower *Lou* 樓 in the front. Sima Qian mentions *tian ku lou* 天庫樓 including many stars in the southern part of the *xiu Zhen* 軫. The 29 stars include five sets of poles, *Zhu* 柱 to bind horses and a place for drilling and examining soldiers, *Heng* 衡.

64. *Nanmen* 南門; ξ_2 (Cen), 2 stars;
the South Gate of the arsenal. These two stars were six degrees above the southern horizon when culminating in the autumn.

65. *Ping* 平; γ (Hya), 2 stars;
a flat plain, good for driving a carriage.

66. *Qiguan* 騎官; c_1 (Cen), 27 stars;
cavalry officers.

67. *Jizu* 積卒; θ (Lup), 12 stars;
many soldiers ready for battle, in groups of three. During the Qin and the Han, continuous military actions took place against the invading Xiongnü 匈奴. This culminated during the reign of Han Wu Di. More in Appendix II, sub. *Cheqi* 車騎, nr. 81 of GS.

68. *Gui* 龜; GC 24374, 5 stars;
a small tortoise, fit for human consumption. The sun arrived at this position at the end of the autumn. Then the emperor went to the *yue tan* 月壇 temple in the western part of the capital to bring his offerings to Hou Ji 后稷 and Huangdi.

The Greek poet Aratos (c. 250 BC) reported on "Ara", an Altar in the sky at the same position as the Chinese *Gui* 龜.

69. *Fuyue* 傅說; G (Sco), 1 star;
a legendary wise minister. According to Sima Qian, Fu Yue was a chancellor of Emperor Wu Ding of the Yin dynasty. It could also represent a shaman in the royal harem helping to produce sons.

70. *Yu* 魚; M7 (Sco), 1 star;
fish roe, symbol of prosperity.

71. *Chu* 杵; GC 25094, 3 stars;
a pestle to prepare food (on the Altar). See sub. *Gui*, nr. 68.

72. *Bie* 鱉; μ (CrA), 14 stars;
a big river turtle with a soft belly and a strong shell. In Chinese mythology the three Islands of the Blessed in the Eastern Sea rest on the heads of three huge turtles.

73. *Jiukan* 九坎; ι (Sgr), 9 stars;
nine water wells, representing the irrigation system in the field. These wells are connected with underground channels to distribute water. This method of irrigation has been used in Xinjiang for more then two thousand years.

74. *Baijiu* 敗臼; ι (PsA), 4 stars;
a tub shaped like a large mortar for waste disposal.

75. *Yulinjun* 羽林軍 and *Leibizhen* 壘壁陣; υ (Aqr), 57 stars;
the *Yulinjun* were Royal Guards, big soldiers drafted from the Northern Territories. The *KYZJ* called them the army of the northern barbarian kings.
Leibizhen is a row of fortifications and trenches protecting the winter camp.

76. *Beiluo shimen* 北落師門; α (PsA), 1 star;
the South Gate of the camp of the *Yulinjun* in *bei luo* 北落, the "Northern Territories." Northern Territories were synonymous with the land of darkness and void. In the sky it was in the winter quarter.

77. *Tusikong* 土司空; β (Cet), 1 star;
Master of Constructions and Earth-works.

78. *Tiancang* 天倉; η (Cet), 6 stars;
a barn for grain.

79. *Tianjun* 天囷; α (Cet), 13 stars;
a granary for corn (maize).

80. *Tianlin* 天廪; o (Tau), 4 stars;
a granary for rice.

The Zhou rulers instructed the farmers to build granaries for the preservation of their crops.

81. *Tianyuan* 天苑; γ (Eri), 16 stars;
celestial meadows for small cattle to be sacrificed to the Gods.

82. *Shenqi* 参旗; π_5 (Ori), 9 stars;
the banner of *Shen*.

83. *Yujing* 玉井; β (Eri), 4 stars;
the jade well, only to be used by nobility.

84. *Ping*屏; μ (Lep),2 stars;
the screen for privacy of the people using the toilet *Ce* 廁.

85. *Ce* 廁; β (Lep), 4 stars;
a toilet with a shed, according to Sima Qian.

86. *Tianshi* 天屎; ν_2 (Col), 1 star;
excrement from the toilet *Ce*.
Sima Qian says: "Yellow excrements are a good sign, but green, white or black is fatal." Excrements of men and animals provided fertilizer for the fields. They were a part of the agricultural eco-system.

87. *Junshi* 軍市; 17 (Lep), 13 stars;
a market for the soldiers.

88. *Yeji* 野雞; β (CMa), 1 star;
a pheasant.

89. *Lang* 狼; α (CMa), 1 star;
a wolf.

90. *Hu* 弧; κ (CMa), 9 stars;
the bow. Sima Qian calls it *Hushi* 弧矢, bow and arrow. He says that the arrow *Shi* is directed at the wolf *Lang* 狼.

91. *Laoren* 老人; α (Car), 1 star;
the wise, old man. Sima Qian called him *Nanji laoren* 南极老人, the Old Man of the South Pole, who brought peace and happiness to everyone.

A picture of the Ploughing Ceremony. This ceremony was performed by ancient emperors to encourage agriculture. It happened in mid-spring, the season for sowing. The holy field was represented in the sky by the constellation *Tiantian* (Celestial Field) near *xiu Jiao* (α Vir), where the spring equinox was located during the Han.

Photo taken from Berliner Festspiele Katalog, *Europa und die Kaiser Chinas* (Insel Verlag; 1985).

A picture of the autumn harvest, from the *Peiwen gengzhi tu* (imperial edition of 1696 AD). The poem accompanying the picture reads as follows:

Harvest Poem

High time for the farmers to reap;
Sickles wave faster and faster, the rice fills the barn.
Severe cold chaps their hands like a tortoise shell;
All day they are working with their bent back.
Children are busy collecting scattered ears;
Cold wind penetrates their short, coarse clothes.
Happily they carry their harvest home,
With a full and friendly moon above the hills.

On the Lunar Calendar this picture corresponded to the *ci* (Jupiter's Mansion) *Jianglou* during the Han. *Jiang Lou* meant ''Descendant of the Reaper'' and referred to specific constellations of *Shi Shi*: *Tiancang* (barn), *Tianjun* (granary for corn), *Tianlin* (granary for rice) and *xiu Lou* (sickle of the reaper). See Appendix I of this book.

During the Han the *Laoren* star appeared just above the southern horizon of Chang'an during two hours. It culminated after sunset about mid-autumn. (*TGS*).

92. *Ji* 稷; δ (Vel), 5 stars;
a temple or shrine to worship Hou Ji 后稷. In the last two weeks of the summer season *da shu jie* 大暑節 (22 July to 7 August) there were big festivals with sacrifices to Hou Ji 后稷, Huangdi 黄帝 and other important Ancestors. At that time the solar god *Xuanyuan* passed from the *xiu Xing* 星 into the *xiu Zhang* 張 which corresponded exactly with the celestial position of *Ji* 稷, low above the southern horizon. The temple ceremonies were described in *Jin shu.*

---

*The following are the names of the twenty eight lunar lodges (*xiu*) with some brief explanations.*

93. *Jiao* 角; α (Vir), 2 stars;
the horn of the Green Dragon *cang long* 蒼龍, where the main gate to the palace complex of *Xin xiu* is located. See 6.3.3.

94. *Kang* 亢; κ (Vir), 4 stars;
the neck of *cang long;*
outer palace of the emperor for daily business.

95. *Di* 氐; α (Lib), 4 stars;
the paw of *cang long;*
palace for the night, for the emperor and his wife plus two important mistresses.

96. *Fang* 房 and *Gouqian* 鉤鈐; π (Sco), 4 plus 2 stars;
the breast of *cang long*;
the place where the emperor announced his policy for the coming year (*Mingtang* 明堂). In that case *Gouqian* is the key of the doorway to the palace.

97. *Xin* 心; σ (Sco), 3 stars;
the heart of *cang long;*
seat of the emperor with a prince on both sides in the Bright Hall (*Mingtang* 明堂).

98. *Wei* 尾; μ (Sco),9 stars;
the tail of *cang long;*
palace in the rear; the harem, in which many princes live.

99. *Ji* 箕; γ (Sgr),4 stars;
a winnow basket to separate grain from chaff;
Sima Qian also calls it a place of slander behind the back palace.

100. *Dou* 斗; ϕ (Sgr), 6 stars;
the southern dipper *Nandou* 南斗.

101. *Niu* 牛; β (Cap),6 stars;
the ox.
Sima Qian says here the ox is sacrificed for the midwinter ceremonies.

102. *Nü* 女; ϵ (Aqr), 4 stars;
a woman.
Sima Qian calls her a maid, serving everybody and making and repairing everything.

103. *Xu* 虛; β (Aqr), 2 stars;
emptiness, darkness.
Sima Qian calls it the Northland, a place for winter ceremonies and for mourning the ancestors.

104. *Wei* 危 and *Fenmu* 墳墓; α (Aqr), 3 plus 4 stars;
the roof of a house; and the tomb.

105. *Shi* 室 and *Ligong* 離宮; α (Peg), 2 plus 6 stars;
a temple or shrine and summer palaces of the emperor.
Jin shu says here is a place for contemplation and offerings to the ancestors. But here is also a granary for the army.

106. *Bi* 壁; γ (Peg), 2 stars;
the eastern wall of the temple grounds.
Jin shu calls it a secret library for court officials.

107. *Kui* 奎; ζ (And), 16 stars;
a swine.

In *Tianguan shu, Han shu* and *Jin shu, Kui* is called *fengshi* 封豕, the "Greedy Swine".

108. *Lou* 婁; β (Ari), 3 stars;
the junction point or the sickle of a reaper to cut the harvest.

According to Sima Qian, *Lou* is the place of "gathering" (*ju zhong* 聚眾) animals (cows or sheep) for sacrifices. During the Han the full Moon at β (Ari) corresponded with the autumn equinox. One month later the emperor offered a cow or a ram at the Temple of the Moon, *yue tan* 月壇. Therefore the *Jin shu* said that *xiu Lou* 婁 was a field with animals to be sacrificed. That field was *Tianyuan* 天苑, nr. 81.

109. *Wei* 胃; 35 (Ari), 3 stars;
thick stomach; fat belly[1].
Both *Tianguan shu* and *Jin shu* say here are storehouses (granaries) for grain. Indeed in *xiu Lou* and *xiu Wei* 胃 we find *Tianjun* 天囷 and *Tianlin* 天廩, nrs. 79 and 80.

110. *Mao* 昴; 17 (Tau), 7 stars;
lodge of abundance or of the setting sun.
The full Moon in *Mao* meant abundance of food. That was the time of the traditional sacrifices to Huangdi.

Sima Qian said *xiu Mao* is the lodge of white clothes for mourning the dead. On tombs the full Moon was always depicted in *xiu Mao* as a sign of death.

111. *Bi* 畢 and *Fu'er* 附耳; ϵ (Tau), 8 plus 1 stars;
a hunting net to catch animals. *Fu'er*, the "ear", may represent the man who hears animals coming.
The whole scene depicts the *qiu xian* 秋獻 (autumn sacrifice) hunting ceremony performed by the Emperor at the *li dong jie* 立冬節 (middle of November).
Sima Qian describes a hunters' carriage with arrows and lances. *Jin shu* describes a complete funnel formed with many soldiers along its sides to prevent the flushed game from escaping. At the end of this hunting trap *Bi* 畢 the emperor is waiting to shoot the animal with bow and arrow, *Hushi* 弧矢, nr.90.

112. *Zui* 觜; ϕ (Ori), 3 stars;
the beak of a bird.
In hunting trained animals were used to rouse deer and smaller animals. Trained falcons were used to catch rabbits. Probably the falcon

[1]*Xiu Wei* 胃 corresponds with the Hindu lunar lodge Bharani standing for a thick belly. See E. Burgess [1935].

was the hunting-bird indicated by *Zui*. Sima Qian suggested *Zui* was the head of the Tiger *bai hu*.
Another suggestion by Sima Qian and in the *Jin shu* was that *Zui* is a commander in charge of provisions for the troop, which in fact had the same function of providing food, just like the hunting party did.

113. *Shen* 參; δ (Ori), 10 stars;
the big warrior or hunter.
Sima Qian said *Shen* was the White Tiger *bai hu* 白虎. In November the tigers descended from the mountains. On the other hand a Warrior was given this name in a metaphoric way to indicate his bravery and strength. Sima Qian also speaks about the left and right shoulders of the warrior, clearly meaning α and γ Orionis. *Shenqi* 參旗 was his royal standard, and *Fa* 伐 his dagger according *Shi Shi*.
Other meanings mentioned in the *Jin shu* as well as by Sima Qian were
— a big army camp with several generals, or
— an autumn market with the central Triaster as the weighing beam.

114. *Jing* 井; μ (Gem), 8 stars;
the eastern well.

115. *Gui* 鬼; θ (Cnc), 5 stars;
ghosts in *Yugui* 輿鬼 (ghost wagon), according to the *Jin shu* and other sources.
The milky spot Praesepe (ϵ Cnc) visible in *Gui* is an open star cluster. All ancient civilisations noticed this spot and had mystical associations with it, like millions of souls waiting to be incarnated.

116. *Liu* 柳; δ (Hya), 8 stars;
a willow.
The willow is the Chinese symbol of Life. Dead bodies were wrapped with willow twigs. This picture corresponded with the lunar calendar, at midwinter. But on the solar calendar *xiu Liu* corresponded with the first half of July, *xiao shu jie* 小暑節 during the Han. Therefore *Jin shu* mentioned the big activities of the field kitchen *Waichu* 外廚 to serve the guests at the summer festivals. For *Waichu* see Appendix II, nr.115.

Sima Qian called *xiu Liu* the lodge of plants and trees and the beak of the Red Bird, *zhu que* 朱雀.

117. *Xing* 星; α (Hya), 7 stars;
the bird star.
This is one of the four cardinal asterisms mentioned in the *Yaodian* 堯典. It is synonymous with mid-summer. According to the *Tianguan shu* the constellation *Xing* 星 is the neck of the Red Bird *zhu que.*
The *Jin shu* says the celestial city where people wear beautiful clothes with flowers is here. The star *Niao,* α (Hya) is located within one degree of the celestial equator at the time of Yao Di. This equatorial position of the *ju xing* of *xiu Xing* 星, together with the following *ju xing* of *xiu Zhang* and *xiu Yi* is a strong point to demonstrate the existence of an independent Chinese system of Lunar Lodges. The corresponding Indian and Arabic determinative "stars" follow the ecliptic, about 30 degrees more to the North.

118. *Zhang* 張; υ (Hya), 6 stars;
a drawn bow or a bird catcher.
The red lacquered bow of the Archer Yi. The *Tianguan shu* mentioned it as the stomach of the bird.

119. *Yi* 翼; α (Crt), 22 stars;
a wing to fly on the wind.
The wings of the Red Bird *zhu que*.

120. *Zhen* 軫; γ (CrV), 4 stars;
the rear of an ancient carriage.
The *Jin shu* calls *Zhen* a place of death which corresponds with the mausoleum *Daling* 大陵 (nr. 34) on the opposite side of the sky.
We want to mention that at this position on the Sumerian Sky is a Crow named UGA, probably identical with the three-legged crow *san zu wu* 三足烏, representing the Sun in Chinese mythology. UGA became the Roman crow Corvus, which is now the western name for the same constellation. See Papke [1989].

A picture from the *Tiangong kaiwu* (by Song Yingxing of the Ming dynasty; 1637). It originally illustrated the use of the lever as a mechanical device. Here we use it to illustrate the constellation of *Jing* (well), which in shape resembles Chinese character 井 on the well

APPENDIX II

THE CONSTELLATIONS OF GAN SHI

In this appendix we present in detail a historical approach to each constellation of *Gan Shi*. The constellations are listed in the same sequence as in the *Kaiyuan zhanjing*. Each entry gives particulars in the following order:

1. The star name in *pinyin* spelling, together with Chinese characters.
2. The identification of one star in the constellation. This star is our foot-hold for the reconstruction of the constellation and also serves as a guide to our star maps. The coordinates of these guiding stars are given in historical catalogues. The most important ones are

   - The star catalogue observed by Zhou Cong 周琮 in the Huangyou 皇祐 period (1052 AD) of the Song, abbreviated as HY.
   - The star catalogue in the *Tianwen huichao* 天文彙鈔 observed at the beginning of the Ming (1375 AD), abbreviated as TW.

3. The relative position and the number of stars in the constellation. Primarily they are given according to *Kaiyuan zhanjing*. If not, we will give special notice.
4. A brief translation or circumscription of the name of the constellations.
5. Explanations and comments. We tried to understand each constellation in the context of the sky and against the historical and cultural background which provides valuable inferences for our approach.

By "(SS)", "(GS)" or "(WS)" after a star name we indicate the school to which this constellation belonged: *Shi Shi*, *Gan Shi* or *Wuxian Shi*.

1. *Tianhuang dadi* 天皇大帝 (? UMi), 1 star inside *Gouchen* (SS); High God of Heaven.

   Before the Zhou dynasty *Tian* 天 and *Di* 帝 had the same meaning. Only after the Zhanguo period *Huang Di* 皇帝 was used in combination with *Tian* 天 to indicate the ultimate Heaven. During the Han, due to the growing influence of Taoism, the title *Tianhuang dadi* was sanctified, meaning the ultimate deity in the Taoist hierarchy, ruling in the office at the Pole, *Beiji* 北极.

2. *Sifu* 四輔 (? UMa), 4 stars surrounding the North Pole; four advisors of the Emperor.

   There were four consultants surrounding the Emperor, according to an ancient Zhou tradition. Their titles were *Ni* 疑 (for general consults), *Cheng* 丞 (for administrative consults), *Fu* 輔 (for civil consults) and *Bi* 弼 (for military consults). The *Ni* sat in front, *Cheng* behind, *Fu* to the left and *Bi* to the right of the Emperor. This was a typical example of Confucian symbolism.

   When Wang Mang took the position of the Royal Han family of Liu, establishing the short-lived Xin dynasty, he actually brought this system of four advisors into practice under the names:

   tai shi 太師, the Superior Teacher;
   tai fu 太傅, the Superior Master;
   guo shi 國師, the State Teacher;
   guo jiang 國將, the State General.

3. *Huagai* 華蓋 and *Gang* 杠 (? Cas), 7 plus 9 stars, above *Tianhuang dadi* (GS);
   the canopy of the emperor.

4. *Wudi neizuo* 五帝内座 (γ Cep?), 5 stars inside *Ziwei* (SS) below *Hua Gai* (GS);
   inner seats of Five *Di*.

   Introduced by *Gan Shi* to distinguish it from *Wudizuo* 五帝座 of *Shi Shi* in *Taiwei*. The central court was more essential than any other court in the sky; therefore the Five *Di* should have their seats in the Inner Court.

5. *Liujia* 六甲 (? Cep), 6 stars inside *Ziwei*, near the handle *Gang* 杠 of the *Huagai*;
   the Six *jia* 甲.

   In ancient China, a sexagesimal numbering system was used in calendars. A series of ten stems, *gan* 干 were combined alternatively with a series of twelve branches, *zhi* 支 so as to make sixty combinations. This sexagesimal system can be thought of in the image of two enmeshed cogwheels, one having twelve and the other ten teeth, so that not until sixty combinations have been made will the cycle repeat. It can also be presented in a 6x10 matrix, six rows of ten stems are matched alternatively with the twelve branches, five series of the branches being needed to complete the match. If we use alphabets to denote the ten stems and twelve branches as following:

A	B	C	D	E	F	G	H	I	J
甲	乙	丙	丁	戊	己	庚	辛	壬	癸

and

a	b	c	d	e	f	g	h	i	j	k	l
子	丑	寅	卯	辰	巳	戊	未	申	酉	戌	亥

then the sexagesimal matrix is as follows:

1	Aa	Bb	Cc	Dd	Ee	Ff	Gg	Hh	Ii	Jj	10
11	Ak	Bl	Ca	Db	Ec	Fd	Ge	Hf	Ig	Jh	20
21	Ai	Bj	Ck	Dl	Ea	Fb	Gc	Hd	Ie	Jf	30
31	Ag	Bh	Ci	Dj	Ek	Fl	Ga	Hb	Ic	Jd	40
41	Ae	Bf	Cg	Dh	Ei	Fj	Gk	Hl	Ia	Jb	50
51	Ac	Bd	Ce	Df	Eg	Fh	Gi	Hj	Ik	Jl	60

   Jia (A) was the first of the ten stem, thus in this matrix each row was headed by *jia*. Hence the Six *jia (Liujia).*

   During and after the Zhanguo period magic numbers played an important role in philosophy. Astrologers used the *gan zhi* to make divination tables. During the Han several kinds of divination systems were used on diviner's boards in which the *gan zhi* was incorporated. One of them was called *qi men dun jia* 奇門遁甲 in which the Six *Jia* (*Liujia*) were mysthicized and had a strong influence on the final divination outcome. Apparently these Six *Jia* were meant by *Gan Shi.*

6. *Tianzhu* 天柱 (? Dra), 5 stars inside *Ziwei* close to the eastern wall; pillars for royal announcements.

7. *Zhuxiashi* 柱下史 (χ Dra, HY), 1 star inside *Ziwei* North-East of *Beiji;*
   officer in charge of public announcements, recording what the Emperor said.
8. *Nüshi* 女史 (ψ Dra, HY), 1 star North of *Zhuxiashi;*
   woman officer in charge of the waterclocks in the royal palace.
9. *Shangshu* 尚書 (15 Dra, HY), 5 stars in the South-East of *Ziwei;*
   master of writing, secretary.

   These officials were introduced for the first time during the Qin dynasty. Qin Shi Huangdi had four of them at court in charge of his administration.

   During the reign of the first rulers of the Former Han, especially during the reign of Emperor Wu, these positions were taken by eunuchs to be sure of their trustworthiness. Only during the reign of Emperor Cheng 成, in the fourth year of Jianshi (29 BC) the secretariat was reorganised to a full-shape administration center with a chief secretary plus four heads of secretarial departments. Apparently this was the *Shangshu* projected by *Gan Shi*. The number of stars being five and not four, suggests that the naming of this constellation by *Gan Shi* happened after 29 BC. It is an important indication of the time when this school was active.
10. *Yinde* 陰德 (? UMi), 2 stars West of *Shangshu;*
   virtue of the *yin*, hidden virtue or secret power.

   This Taoist indication of secret power concerns the invisible power (*qi* 氣), exerted by *Taiyi* or *Tianyi* from their high position (7, 8 Dra). These two celestial positions for *Tianyi* and *Taiyi* were indicated certainly by *Shi Shi.*

   Sima Qian mentioned *Yinde* together with *Tianyi*, saying

   > In front of the North Pole Office *Beiji*, facing the `opening' of *Beidou* are three stars shaped like a cone with its point to the North. Sometimes they are visible, sometimes not. They are called *Yinde* or *Tianyi.*

   The three stars mentioned were 7, 8, 9 Dra. Probably the word "*or*" in "They are called *Yinde or Tianyi*" should be "*and*". Then one of *Tianyi* plus two of *Yinde* makes three. Furthermore *Jin shu* mentioned *Yinde* and *Yangde* 陽德 (Virtue of the *yang*) instead of *Yinde*. So probably the two characters of *Yangde* (virtue of the *yang*) were missing in the *Tianguan shu*. In that case *yin* plus *yang* plus *yi* (One) makes three, which would fit the *yin-yang* philosophy quite well: *yin* and *yang* are the two halves of *Tianyi* , the Heavenly One. We discovered some evidence for the existence of *Yinde* and *Yangde* instead of only *Yinde* alone. On the *Suzhou tianwen tu* from the Song dynasty *Yinde* and *Yangde* were clearly indicated as two constellations, each with one star.

But the above explanation raises another question: How about *Taiyi* of *Shi Shi*? We suggest that *Taiyi* originally was the same as *Tianyi* or *Dayi* 大一 (Great One), because the Chinese characters for *Tai* 太, *Tian* 天 and *Da* 大 were of common origin and meaning and were often used interchangeably before the Han. It was Taoist philosophy, seeking the ultimate *Dao* 道, that distinguished *Taiyi* from *Tianyi* and endowed the former with a still higher attribute. Both *Tianyi* and *Taiyi* could mean equally God of Heaven before the Han. But during the reign of Emperor Wu a scholar Miu Ji 謬忌 claimed that "*Taiyi* is the noblest God of all Heavenly Gods." He suggested a special altar be built for sacrifices to *Taiyi*, which was all agreed by Emperor Wu. See *Shi ji fengchan shu* 史記 • 封禪書. So *Taiyi* became superior to *Tianyi* and was the origin of *Yinde* and *Yangde*. Influenced by philosophical ideas of the time, *Shi Shi* indicated both *Taiyi* and *Tianyi* among those three stars. Then there was no more space for the two stars of *Yinde* and *Yangde*. We suppose that this was the reason that *Gan Shi* moved these two constellations into the *Ziwei Yuan*.

11. *Tianchuang* 天床 (? Dra), 6 stars "outside the door of *zi gong* 紫宮 (the Purple Palace, i.e., *Ziwei*);"
celestial bench or mattress.

12. *Tianli* 天理 (? UMa), 4 stars inside the "scoop" of *Beidou;*
the Great Judge for nobility.

The *Tianguan shu* described the "scoop" of *Beidou* as a prison for criminals of a noble rank. *Shi Shi* denoted a Celestial Prison *Tianlao*

天牢 right beneath the "scoop" of *Bei Dou*; while *Gan Shi* projected the Great Judge *Tianli* 天理 in the "scoop" of *Beidou*.

13. *Neichu* 内廚 (4 Dra?), 2 stars at the South-West corner of *zi gong;* the inner kitchen.

    There were three kitchens in the sky, all according to *Gan Shi*
    — *Tianchu* 天廚, the celestial kitchen outside *Ziwei*
    — *Neichu* 内廚, the inner kitchen for royalty only
    — *Waichu* 外廚, the outer kitchen in the southern sky (*wai guan* 外官) for common use.

14. *Neijie* 内階 (2 Dra), 6 stars to the North of *Wenchang* (SS); the inner steps.

    To pass from *Taiwei* to *Ziwei* one had to use the outer steps, *Santai* indicated by *Shi Shi* to reach *Wenchang*. The three inner steps meant by *Gan Shi* lead further from *Wenchang* to *Ziwei*.

15. *Tianchu* 天廚 (δ Dra, HY), 5 stars outside the North-Eastern wall of *Ziwei*;
    the celestial kitchen. See *Neichu*.

16. *Ce* 策 (κ Cas), 1 star in front of *Wangliang* (SS);
    the whip of Wang Liang 王良, the legendary charioteer of the State Jin.

17. *Chuanshe* 傳舍 (GC3947 Cam, HY), 9 stars "above" *Huagai*, alongside the Celestial River;
    the guest rooms.

    Residential palaces for guests from remote countries or provinces.

18. *Zhaofu* 造父 (δ Cep, HY), 5 stars South of *Chuanshe,* in the middle of the River;

    Zhao Fu, the legendary charioteer of Mu Wang 穆王, Great King of the Zhou (1001 - 946 BC).

    The legend tells us that while Mu Wang was visiting Xi Wang Mu 西王母 in the far west, a rebel rose against the Zhou in the province of Zhao. In this emergency his charioteer drove him back just in time to suppress the revolt. Mu Wang gave the province to his excellent driver who got the name Zhao Fu 造父, the ancestor of the Zhao 趙 state.

19. *Chefu* 車府 (72 Cyg, HY), 7 stars East of *Tianjin* (SS), near the River;
a big yard for chariots.
20. *Ren* 人 (κ Peg, HY), 5 stars South-East of *Chefu;*
a human being.

The *Tianguan Shu* says that alongside of *Wangliang* there are eight stars indicating a shallow place in the celestial river. Sima Qian called them *Tianhuang* 天潢 (*Shi Shi* called them *Tianjin* 天津). If the stars in the Celestial River *Jiang* 江 twinkle, "*Ren* 人 (people) are wading across the river." Possibly *Gan Shi*'s *Ren* was derived from this astrological protasis[1].

21. *Neichu* 内杵 (π Peg, HY), 3 stars near *Ren;*
the inner pestle.
22. *Neijiu* 内臼 (ι Peg, HY), 4 stars South-East of the nearby *Ren;*
the inner mortar.

The presence of the pestle *Chu* 杵 and the mortar *Jiu* 臼 near *xiu Shi* 室 implies the preparation of a meal. On the same hour-circle is *Tianchu* 天廚 near *Ziwei*. According to the *Tianguan Shu*, *xiu Shi* and *Bi* 壁 are places to worship ancestors (*qing miao* 清廟). Cooking facilities were needed to make special food to be sacrificed to the ancestors.

23. *Fukuang* 扶筐 (39 Dra, HY), 7 stars North of *Tianjin* (SS);
a basket for mulberry leaves.

A basket to be carried by women indicating the ancient Chinese silk-worm culture. This constellation is related to the Spinning Damsel *Zhinü* 織女. It culminated at midnight in June, the high season for silk-worm breeding.

24. *Siming* 司命 (25 Aqr, HY), 2 stars North of *xiu Xu*;
a half-deified judge of fate, deciding on life and death.

According to the study by *Wen Yiduo* [1934] this deity was in charge of the Dark World of Death *xuan ming* 玄冥, which was also associated with "North." *Siming* was mentioned already in Qu Yuan's

[1] It is remarkable that in the Sumerian sky, at the same celestial position is a dead human body named LU · (USH). See Papke [1989].

poetry *Li shao* 離騷 (Suffering Sorrow) from about 250 BC at the end of the Zhanguo period. The constellation *Siming* consisted of two stars near *xiu Xu*, the so called "Northern Palace," where the sun was in mid-winter during the Zhou.

25. *Silu* 司祿 (4 Peg, HY), 2 stars North of *Siming;*
Judge deciding about honours (titles, positions and salaries).
26. *Siwei* 司危 (β Equ, TW), 2 stars North of *Silu*;
Judge deciding about cowardice and bravery.
27. *Sifei* 司非 (γ Equ, TW), 2 stars North of *Siwei*;
Judge about deviations from cosmic harmony.

In *Gan Shi*'s sky there are five star names with the component *Si* 司 which means " in charge of." These celestial projections represent just a few out of many purely imaginary deities from the Han time. The exalted titles of these imaginary functionaries had in general no counterparts in the secular officialdom of the Han, except in a few cases which we have indicated in the table on the next page.

28. *Baigua* 敗瓜 (ϵ Del, HY), 5 stars near *Hugua* (SS);
an over-ripe melon.
29. *Zuoqi* 左旗 (γ Sge, HY), 9 stars near *Hegu* (SS);
a banner on the left side of *Hegu*.
30. *Tianji* 天雞 (55 Sgr, HY), 2 stars North of *Gouguo* (SS);
a celestial cock.
31. *Luoyan* 羅堰(τ Cap, HY), 3 stars East of *xiu Niu*;
a network of dikes for agricultural purposes. See also *Tiantian* the celestial fields; sub II, 86.
32. *Shilou* 市樓 (μ Oph, TW), 6 stars above *xiu Ji* and inside the *Tianshi;*
the market tower.
33. *Hu* 斛 (28 Her, TW), 4 stars South of *Dou* (SS);
a square box for measuring, probably a standard calibre.

The above two constellations had a function in the Celestial Market *Tianshi.*

34. *Ri* 日 (2 Sco, HY), 1 star "in front of the middle door-way of *xiu Fang*;"
the Sun star.

Deified star names	Correlated secular Officials
Shangjiang 上將, Superior General	*da jiangjun* 大將軍, Great General representing the Greatness of Military Power
Cijiang 次將, Junior General	*shang shu* 尚書, Master of Writing, secretary in charge of civil administration
Guixiang 貴相, Honourable prime minister	*tai chang* 太常, Grand Master of Ceremonies
Silu 司祿, in charge of ranks *Sizhong* 司中, in charge of positions	*si li* 司隸, minister of labour
Siming 司命, in charge of fate *Siguai* 司怪, in charge of omens	*taishi ling* 太史令 Prefect of Astrologers to eliminate harm by abnormal phenomena
Sikou 司寇 in charge of law and order	*da li* 大理 High Judge of Court

Table 8.1: A survey of imaginary star names connected with officialdom. On the left are imaginary deified star names; on the right are corresponding functions or positions in secular officialdom.

35. *Tianru* 天乳 (μ Ser, HY), 1 star North of *xiu Di*;
 the celestial nipple; a breast producing milk.
36. *Kangchi* 亢池 (μ Vir?), 6 stars North of *xiu Di;*
 a lake belonging to *xiu Kang*.

 According to the *Sui shu*, on this lake ceremonies were held to welcome foreign guests arriving by boat. In that case *kang* should be pronounced as *hang* which means sailing; and the name of the constellation could be translated as "Sailing Waters."

 Also according to the *Huangdi zhan* quoted in the *KYZJ Kangchi* was a ferry to cross the water. Another name was *Kangxing* 亢星, Stars of *Kang* 亢.

 The *KYZJ* quoted *Gan Shi* saying: " The six stars of *Kangchi* are to the North of *xiu Kang*." The same six stars we see on the Song map, but very close to *Dajiao*, to its South. At the same position the Qing catalogue *Yixiang kaocheng* also indicated *Kangchi* but with four very faint stars. On the contrary, the *Huangdi Zhan* mentioned bright stars in *Kangchi*.

 The above suggests that during the Han *Kangchi* might have been at a completely different position in the sky as indicated on later maps. Looking at our star map we suggest that *Gan Shi* meant six bright stars on the border between *xiu Kang* and *xiu Di*, namely 110, 109 and μ Vir, and 11,16 and δ Lib. Another argument for our inference can be obtained from the description about the position of *Tianru* 天乳 by *Gan Shi:* "*Tianru* is facing the `opening' of *Kangchi*." The position of *Tianru* is certainly μ Ser. How could the description of *Tianru* be true if *Kangchi* did not mean the six stars which we suggest?
37. *Jiantai* 漸臺 (γ Lyr, HY), 4 stars attached to the Eastern "foot" of *Zhinü* (SS);
 a waterside pavilion.

 Jin shu and *GSXJ* said that *Jiantai* presided over music, sundial and waterclock. It symbolized the unification of standards in China for Tones of Music, for Weight, Area, and Length and for the measurement of Time. *Jiantai* was the Office where Standards were kept, like in a modern Bureau of Standards. According to the *yin-yang* philosophy, during mid-winter the cosmic wind *qi* had maximum influence. Then was the right time to calibrate standards for everything.

Jiantai had an installation with twelve calibrated bamboo-tubes used for the tuning of the royal musical instruments[2].

The base tone from which all other tones were derived, was given by the Yellow Bell (*huang zhong* 黄鍾) and corresponded with the tone generated by the *qi* in an open bamboo-tube of 9 *chun* 寸 long and 3 *fen* 分 wide. It was strongest in the 12th month of the year.

The corresponding length of a tempered chord on the imperial *zhun* 準 (string instrument) provided the unit-length introduced by Qin Shi Huangdi. From this length the standards for area, volume and weight were derived. See *Hou han shu luli zhi* and Joseph Needham [1959].

The introduction of these standard measures on the autumn market is visible west of the Midwinter hour circle, in *Tian Shi*, where the *Dou* (SS), the *Hu*(GS) and the *Bodu* (WS) are in use.

During the Han the hour-circle passing through *Jiantai* coincided with the position of the sun on the ecliptic at mid-winter.

38. *Niandao* 輦道 (13 Lyr, HY), 5 stars attached to the western "foot" of *Zhinü*[3];
a carriage road connecting royal palaces.

39. *Sangong* 三公 (24 CVn, HY), 3 stars South of the handle of *Beidou;*
Three Excellencies, direct advisors of the celestial Emperor.

40. *Zhouding* 周鼎 (6 Boo, HY), 3 stars West of *Sheti* (SS);
Tripod of the Zhou, a caldron.

The Nine Tripods of the Zhou were regarded as National Emblems. When the Zhou dynasty declined these tripods were lost. Qin Shi Huangdi dreamed of retrieving them. According to the *Shi ji* he ordered 4000 men to dive into the river Si 泗 in order to find them, but without success.

---

[2]Cai Yong 蔡邕 (178 AD) described a hermetically closed room in which 12 standard, open-ended pipes were installed in a circle, according to the 12 months of the year. Each pipe had an inclination relative to the vertical, enough to retain some powder of burned reeds dusted above its upper open-end. According to his description each month of the year was characterized by a specific *qi* which excited preferentially one corresponding tube. This could be seen from the reed ashes disappearing from the upper end of the specific tube of the month.

[3]On the star map *Niandao* is to the east of *Zhinü*.

The Nine Tripods of the Zhou became symbols for the Emperor's Dignity. During the Han searching for these tripods was a favourite topic at court, reason why *Zhouding* was projected in the sky, with three stars representing the three feet of the Imperial Tripod.

41. *Dixi* 帝席 (ξ Boo, HY), 3 stars North of *Dajiao* (SS);
a seat or mattress for the Emperor.

42. *Tiantian* 天田 (78 Vir, HY), 2 stars North of the "right" star of *xiu Jiao;*
the celestial fields.

In this case "right" meant on the right side of the path of the sun. One week after the Spring Equinox the emperor ploughed ceremonially on the imperial fields. After this some high officials, followed by invited farmers had to finish the ploughing. It symbolized that the spring sowing was started by the Son of Heaven. This tradition originated from the beginning of intensive farming, introduced during the later Zhou.

43. *Tianmen* 天門 (53 Vir, TW), 2 stars South of the "left" star of *xiu Jiao;*
the celestial gate.

There were three such gates in the sky: *Tianguan* 天關 (SS); *Tianyue* 天籥 (WS) in Ophiuchus and *Tianmen* 天門 (GS) in Virgo. All these three have an astronomical meaning because the ecliptic passes nearby. They provided a passage for Sun , Moon and planets.

There were three more gates in the sky: *Nanmen* 南門 (SS), the Southern Gate of *Kulou* 庫樓 and *Beiluo shimen* 北落師門 (SS), the Gate of the Military Camp in the Northern Territories, and *Junnanmen* 軍南門 (WS), the South gate of the military camp of the celestial Superior General *Tianda jiangjun* 天大將軍.

These two types of gates imply that there were two different ways of considering the sky during the Han. One was as a dynamic sky, based on the sequence of the seasons. In that case the movement of Sun, Moon and planets along the ecliptic had to be considered. The other was as a static sky which was hovering over the terrestrial "kingdom." Therefore the latter gates were more architectural.

44. *Pingdao* 平道 (74 Vir, HY), 2 stars between the two stars of *xiu Jiao;*
the flat road.

Pingdao was between the two stars of *xiu Jiao* where the ecliptic passes. Here was a road for Sun, Moon and planets.

45. *Jinxian* 進賢 (θ Vir, HY), 1 star West of *Pingdao;*
virtuous people being summoned to the court to be nominated, possibly after having passed the court-examinations.

46. *Yezhe* 謁者 (16 Vir, HY), 1 star North of *Zuo zhifa* (η Vir) of *Taiwei;*
an errand officer who introduces visitors to the emperor.

47. *Sangong neizuo* 三公内座 (? Vir), 3 stars North-East of *Yezhe;*
three seats inside *Taiwei* for the three Excellencies.

Gan Shi had already a constellation of *Sangong* near *biao* 杓, the handle of *Beidou*. In *Taiwei* were the "inner" (*nei* 内) seats of *Sangong*.

48. *Jiuqing neizuo* 九卿内座 (ρ Vir, HY), 3 stars North of *Sangong neizuo*;
the inner seats of Nine Ministers.

Sangong and *Jiuqing* belonged to the idealized Zhou officialdom invented by Han philosophers. But even in *Zhou li* 周禮, the book about Zhou Rituals and Officials, *san gong* and *jiu qing* were not mentioned. Only commentators from the Han time talked about *san gong* and *jiu qing* with great enthusiasm.

According to a recent study by Xu Fuguan [1986], *san gong* and *jiu qing* as a system of officialdom were never put into practice before the Han, as some earlier researchers erroneously believed. It was only until Han Wu Di's time that the story of *san gong* and *jiu qing* became a favourite topic of Han philosophers. Following the advice of Dong Zhongshu Han Wu Di announced that Confucianism was the only " State Ideology;" all other philosophical ideas should be refrained from. According to the Han tradition of Confucianism one should follow the examples of the Zhou in rituals and ceremonies as well as in officialdom. Confucius once said: " As to rituals and ceremonies I prefer to follow the Zhou tradition."

But the ancient officialdom of the Zhou was already in oblivion during the Han. So there was plenty of opportunity for philosophers to give their thoughts free rein. The officialdom introduced in the book *Zhou li* contains a lot of imagination by Han philosophers. Some scholars even think that the book *Zhou li* has been forged by Liu Xin to meet the political needs of Wang Mang[4]. During the reign of Wang Mang all rituals, ceremonies and the official hierarchical system should follow "the Zhou tradition." This way of thinking became so natural that *Gan Shi* projected various aspects of social life onto the sky by nominating constellations like *Sangong* and *Jiuqing*.

49. *Nei wuzuhou* 內五諸侯 (6 Com, HY), 5 stars West of *Jiuqing;* five seats inside *Taiwei* for the five Lords of Heaven.
50. *Taizi* 太子 (93 Leo, HY), 1 star North of *Huangdizuo* (SS); the crown-prince.
51. *Congguan* 從官 (92 Leo, HY), 1 star North-West of *Taizi;* a page.
52. *Xinchen* 幸臣 (2 Com, HY), 1 star North-East of *Huangdizuo;* a favourite official of the emperor.
53. *Mingtang* 明堂 (ϕ Leo, HY) 3 stars outside the South-West corner of *Tai Wei;*
The Bright Hall.

Mingtang as a center for imperial administration during the Zhou was a subject of idealization by Han philosophers. They even invented various names for *mingtang* at different historical periods[5]. At the time of

— Huang Di it was *he gong* 合宮, the Universal Palace;
— Yao Di it was *xu shi* 衢室, the Hall of Passage;
— Shun Di it was *zong zhang* 總章, the General Institution Hall;
— Xia dynasty it was *shi shi* 世室, the Ancestor Temple;
— Yin dynasty it was *yang guan* 陽館, the Hall of *Yang* 陽;
— Zhou dynasty it was *ming tang* 明堂, the Bright Hall.

Advised by Dong Zhongshu (176? - 104? BC) Emperor Wu started to think about a new imperial order by imitating the Zhou tradition. He ordered the building of a *mingtang*. One century later

[4]There were various doctrines and idealizations about the Zhou ascribed to Confucius. Wang Mang used these doctrines to usurp the Han sovereignty of the Liu household. See Gu Jigang [1950].

[5]See the *Tong dian*.

Wang Mang 王莽 was even more interested in *mingtang*. Under the influence of these new philosophies *Gan Shi* created a celestial *Mingtang* outside the Court of *Taiwei*[6].

54. *Lingtai* 靈臺 (59 Leo, HY), 3 stars West of *Mingtang;*
the astronomical observatory; an elevated sacred terrace to observe and read the signs of Heaven, located exactly on the ecliptic.

Like *mingtang* the foundation of the first *lingtai* during the reign of Emperor Wu was a product of the growing influence of Dong Zhongshu's philosophy. His philosophy was a new explanation of the Confucian classic *Chun qiu.* Dong Zhongshu incorporated the ideas of the *yin-yang* philosophies in it and suggested that changing circumstances in the surrounding cosmic world would influence human society. Thus observatories should be built to watch signs of Heaven. The first Han *lingtai* was built eight *li* North-West of Chang'an during Han Wu Di's reign, about 100 BC.

55. *Shi* 勢 (? UMa), 4 stars North of *Taiyangshou* (SS);
a eunuch official, probably it indicated a powerful judge who could decide about punishment by castration.

56. *Neiping* 內平 (30 LMi, HY), 4 stars South of the middle step of *Santai* (SS);
the court of a High Judge who brought people together; a mediator creating equity.

In the Han time the judge was usually a local government official. If he could not solve a problem it went to the provincial governor, and in very difficult cases to the Minister of Justice *ting wei* 廷尉. Perhaps he was the High Judge *Dali* (WS) acting within the inner circle around the imperial court.

57. *Guan* 爟 (χ Cnc, HY), 4 stars North of *xiu Liu* and South of the "tail" of *Xuanyuan* (SS);
a beacon light.

Jin shu describes *Guan* as a watch tower to give light signals in case of military emergencies. The remnants of such towers can be found on the Great Wall even today.

[6]The mean date of appearance of the Full Moon at *Mingtang* was about 14th February, two weeks after the mean date of Lunar New Year, *chun jie* 春節.

58. *Jiuqi* 酒旗 (ξ Leo, HY), 3 stars close to the left horn of *Xuanyuan* (SS);
a tavern-sign to advertise wine shops.
59. *Tianzun* 天樽 (δ Gem, HY), 3 stars North of *xiu Jing*;
a celestial wine cup (with three feet) on the ecliptic.
60. *Zhuwang* 諸王 (τ Tau, HY), 6 stars South of *Wuche* (SS);
several princes.
61. *Siguai* 司怪 (χ^2 Ori, HY), 4 stars in front of *xiu Jing*;
a deity in charge of omina, see Appendix I, no. 112. It could also mean an animal tamer.
62. *Zuoqi* 座旗 (κ Aur, HY), 9 stars North-East of *Siguai;*
a nine starred banner, indicating the presence of the throne of Huangdi.
63. *Tiangao* 天高 (97 Tau, HY), 4 stars close to *xiu Bi;*
a celestial high terrace, perhaps an offering-place.
64. *Lishi* 礪石 (ψ Tau, HY), 4 stars West of *Wuche;*
a whetstone for sharpening knives.
65. *Bagu* 八穀 (β Cam, HY), 8 stars North of *Wuche*;
eight species of grains[7].

    KYZJ mentions rice, broomseed, millet, barley, soybeans, wheat, red beans and sesame seed for these eight grains.
66. *Tianchan* 天讒 (42 Per, HY), 1 star within *Chuanshi* (SS);
a celestial slander or gossip.

    This was added by *Gan Shi* to the curled tongue *Chuanshi* (SS). *Tianchan* and *xiu Mao* were located on the same hour-circle which suggested an interrelation[8].
67. *Jishui* 積水 (λ Per, HY), 1 star within *Tianchuan* (SS);
the accumulated water in the ship *Tianchuan.*
68. *Jishi* 積尸 (π Per, HY), 1 star within the mausoleum *Daling* (SS);
several dead bodies.

---

[7]The occult meaning of the number eight is plenty, abundant like sand. See the translation of *Jin shu* by Ho Pen Yoke [1966]. The star β Cam is in *xiu Mao*, which name means: Lodge of Abundance.

[8]The Chinese *xiu Mao* corresponded to the Indian *Nakṣatra Kṛittika* (7 wives of the 7 wise men Ṛsis). Six of these women had a love affair with Lord *Shiva*, but the seventh (*Arundhati* the Morning-star) refused him. That gave rise to gossip about *Arundhati*. See *Brahmana Purana.*

69. *Zuogeng* 左更 (15 Ari, HY), 5 stars East of *xiu Lou*;
an official in charge of the forest.

The *Sui shu tianwen zhi* mentioned an officer in charge of bamboo in the mountains and of wood in the marshes.

70. *Yougeng* 右更 (107 Psc), 5 stars West of *xiu Lou*;
an official in charge of cows and horses.

Sui shu says *zuo geng* and *you geng* were official posts during the Qin.

71. *Junnanmen* 軍南門 (α Tri, HY), 1 star South of *Tianda jiangjun* (SS);
the south-gate of the army of noblemen.

Gan Shi says that this star is to the South of *Tianda jiangjun*. So it could not be ϕ Andromedae as suggested by Yi Shitong according to the *Yixiang kaocheng*, which is actually to the North of *Tianda jiangjun*. On the Song maps *Junnanmen* was indeed indicated South of *Tianda jiangjun*.

72. *Tianhuang* 天潢 (19 Aur, HY), 5 stars within *Wuche* (SS);
the celestial pond.

73. *Xianchi* 咸池(λ HY), 3 stars within *Wuche;*
the pool of harmony.

Tianhuang and *Xianchi* were projected inside *Wuche* of *Shi Shi*. According to the *Tianguan shu* the five bright stars of *Wuche* were called *Tianwuhuang* 天五潢 (Five Celestial Pond) or *Xianchi*. That meant that the five bright stars of Auriga had several names. When *Shi Shi* had fixed *Wuche* as their name, *Gan Shi* wanted two new constellations inside *Wuche* to convey the ancient meaning of this prominant constellation. *Wuche, Tianhuang* or *Tianwuhuang* and *Xianchi* were all related with the ancient solar mythology. *Huainan zi* said that the Sun, after a day of long driving, bathed in the western pond of harmony *Xianchi*. There might also be a relation with the myth about Dong Huang Gong 東皇公 (the Sun) and Xi Wang Mu 西王母 (the Moon). Dong Huang Gong drove in his solar chariot every day to the western territory of harmony ruled by Xi Wang Mu.

74. *Yue* 月 (37 Tau, HY), 1 star East of *Mao;*
the Moon star. See chapter 6.4.1.

75. *TianJie* 天街 (κ Tau, HY), 2 stars closeby and East of *Yue;*
the celestial street. (for the royal procession to *Tiangao* to bring the autumn sacrifice.)
76. *Tian'e* 天阿 (GC Ari), 1 star West of *xiu Mao;*
a celestial hillside. See also *Tianyin* (WS).
77. *Qingqiu* 青丘 (1 Cen, HY), 7 stars South-East of *xiu Zhen;*
the Green Hills.

In *Jin shu* one reads about *qing qiu*, the green hills in the deep South where barbarians live. By order of Yao Di, the legendary Archer Yi 羿 penetrated into this area to stop the destructive winds coming from there. Yi discovered that *Fei Lian* 飛廉 (or *Feng Bo* 風伯, God of the Wind) was the origin of this misery, because he was offended by other gods. Yi shot an arrow into his knee to bring him to his sense again. See the *Shanhai jing* 山海經 (Classic of Moutains and Rivers), ch.6, 9 and 15.

78. *Zhewei* 折威 (?), 7 stars South of *xiu Kang;*
deprivation of dignity.

Jin shu mentions seven stars of *Zhewei* representing a group of executioners of beheadings in the army.

79. *Zhenche* 陣車 (σ Lib, HY), 3 stars South of *xiu Di*;
battle field chariots.

These chariots were protected in front by leather.

80. *Qizhen jiangjun* 騎陣將軍 (κ Lup, HY), 1 star among *Qiguan* (SS);
the cavalry general.
81. *Cheqi* 車騎 (ζ Lup, HY), 3 stars South of *Qiguan*;
special chariot and cavalry units which were called "garrisons" in the militia system.

This part of the Han sky has been projected on the solar calendar. It corresponded to the end of summer when enough food had been produced by the soldiers to go on expedition. Then the mounted army of Han Wu Di went on war against the Xiongnü.

Between 130 and 119 BC there were annual strikes of Chinese cavalry led by the famous general Huo Qubing 霍去病. This military picture was projected in the sky already by *Shi Shi*. *Gan Shi* described the scenery in more detail by creating some extra constellations.

82. *Kang* 糠 (M7 Sco, TW), 1 star in front of *xiu Ji*;
chaff.

During the Han the sun passed *xiu Ji* 箕 in the second half of November. That was the season to thresh the harvest with flails on the floor. The grains were separated from chaff by shaking a square riddle-basket in the wind. Such a basket was called a *ji* 箕. In November the prevailing wind in central China came from the North-East, reason why the chaff was deposited on the West side of *xiu Ji*.

At the end of the Western Han, Ying Shao 應劭 quoted from the *Li shao* 離騷 saying

> In the autumn *Feng Shi* 風師 helps the farmers to clean the grains. His residence is in *Ji*.

Feng Shi is the god of the wind. See the *Fengshu tongyi* 風俗通義 (ch.8) and the *Huainan zi*.

83. *Nongzhangren* 農丈人 (?), 1 star South-West of *Dou;*
a respected elder farmer, also called husbandman.

Jin shu says he regulates planting and harvesting. *Gan Shi* mentions that he handles the Dipple *Dou* 斗 to measure quantities of grain and works with the winnowing basket *ji* 箕.

84. *Gou* 狗 (51 Sgr), 2 stars "in front of" the scoop of *Dou;*
the dog star.

In this case "in front of " means on the advance side or simply to the East of *Dou*.

85. *Gouguo* 狗國 (ω Sgr, HY), 4 stars South-East of *Jian* (SS);
the territory of dogs around a farm.

86. *Tiantian* 天田 (4 Cap, HY), 9 stars South of *xiu Niu*;
celestial fields.

Imperial fields to be divided by the emperor among provincial governors. Following the Zhou tradition, a ceremony happened in the middle of winter for the renewal of the Fiefs. The emperor took a handful of soil from *Tiantian* and offered it to the attending governor, symbolically renewing his Fief. Probably *Gan Shi* tried to project this tradition into the midwinter part of the sky on the solar calendar.

One should note that *Wuxian Shi* afterwards named a group of constellations called *Shi'erguo* 十二國, the Twelve States next to *Tiantian*. This clarified the picture of this ceremony. During the Zhou all States were vassals to the dynasty. The Fiefs had to be renewed every three years.

87. *Ku* 哭 (ζ Cap, HY), 2 stars South of *xiu Xu;*
the place for mourning or lamentations
88. *Qi* 泣 (ρ Aqr, HY), 2 stars East of *Ku*;
the place where people wept.

Sima Qian called this area the North Land, an area to mourn ancestors. See the *Tianguan shu.*

89. *Gaiwu* 蓋屋 (32 Aqr, HY), 2 stars South of *xiu Wei*;
a roofed house or shed.
90. *Bakui* 八魁 (α Phe, HY), 9 stars South-East of *Beiluo shimen* (SS);
a sway-net catching birds according to the *Jin shu* and the *Kaiyuan zhanjing.*

The birds assemble for migration in early September. On the Lunar Calendar this corresponded with the position of *Bakui.*

On the opposite position in the sky (Solar Calendar) is the *xiu Zhang* which sometimes was also called a sway-net to catch birds (plovers, starlings, ... etc.).

91. *Leidian* 雷電 (σ Peg, HY), 6 stars South-West of *xiu Shi*;
lightning.
92. *Yunyu* 雲雨 (κ Psc, HY), 4 stars South of *Pili* (GS);
cloud and rain.
93. *Pili* 霹靂 (β Psc, HY), 5 stars West of *Tugong* (GS);
thunder clap.

Climatologically the celestial position of the above three constellations corresponded to the two weeks of *da shu jie* 大暑節 during the Han. *Da shu jie* was one of the 24 solar periods of the traditional Chinese farmer's calendar *xia li* 夏曆. It corresponded to the beginning of August. That was the season of storms and lightning especially in the littoral provinces[9].

[9]This group of three constellations, nrs. 91, 91 and 93, belongs to the God of Thunder called *Lei Gong* 雷公. He had five manifestations: his head (or ancestor) *Lei Cu* 雷祖;

Figure 8.1: A picture of farmers separating chaff from grains by means of a winnowing basket; from the *Tiangong kaiwu* 天工開物 (The Use of the Works of Nature) by Song Yingxing 宋應星 of the Ming dynasty. The prevailing wind in the autumn is from the east, so the Chaff, *Kang* 糠 is blown to the west of the Winnowing Basket, *Ji* 箕 (*xiu*), as one can see in the sky.

94. *Tugong* 土公 (ω Psc), 2 stars South of *xiu Bi*;
god of earth works and buildings.
95. *Tugonglii* 土公吏 (ζ Peg, HY), 2 stars South-West of *xiu Shi*;
the assistant of *Tu Gong*[10].
96. *Fuzhi* 鈇鑕 (σ Scl, HY), 5 stars South-West of *Tiancang* (SS);
a sickle to mow the crops.
97. *Tianhun* 天溷 (20 Cet, HY), 7 stars South of *Waiping* (GS);
a celestial sewer near a farm (manure pit).
According to the Qing *Yixiang kaocheng*, *Tianhun* is west of η Cet and much smaller in size.
98. *Waiping* 外屏 (δ Psc, HY), 7 stars South of *xiu Kui;*
the outer fence.

The *Kaiyuan zhanjing* mentions that it is a fence to screen *Tianhun* out of sight.

99. *Tianyu* 天庾 (ν For, HY), 3 stars South-East of *Tiancang* (SS);
ricks of grain in the fields.
100. *Chuhao* 芻藁 (ρ Cet, HY), 6 stars West of *Tianyuan* 天苑 (SS);
medicinal herbs or hay (*Chugao* 芻藁).

Chuhao suggests dried roots of wormwood for chewing. It is a medicine against stomach disease.

101. *Tianyuan* 天園 (v^1 Eri, HY), 13 stars South of the celestial meadows *Tianyuan* (SS);
the celestial orchard.

Tianyuan (GS) was in the same *xiu Bi* as *Xianchi* (GS) and *Wuche*, indicating an astro-mythological connection.

Indeed the *Shanhai jing* mentioned that the Sun ended its daily journey behind the mountain ridge in the West, Yanzi 崦嵫, where

lightning *Dian Mu* 電母; thunder *Lei Gong*; rain *Yu Shi* 雨師 and wind *Feng Bo* 風伯. In the Han sky they are all lined up along one hour-circle and obviously belong together, if we suppose the bright star α Pegasi to correspond to the ancestor *Lei Cu*.

[10]In the Han sky *Tugong* and *Tugongli* were located on the two sides of the group of *Lei Gong*. Apparently they must be a sort of helpers to the God of Thunder *Lei Gong*.

This sector of *xiu Shi* corresponded with the Indian *Nakṣatra Bhadrapada*, in which the same gods were active as in China: *Lei Gong* acted as the Indian *Agni/Indra* (thunder and lightning); whereas *Tugong(li)* as the Indian *Vasus* (earth gods). See Burgess [1935].

the palace of the Goddess Xi Wang Mu and the celestial pond *Xianchi* were located. Moreover Sima Qian confirmed that the 9 stars of *Tianyuan* represented an orchard with peach, apricot and cassia trees, which no doubt was the famous orchard of Xi Wang Mu[11].

102. *Jiuzhou shukou* 九州殊口 (o^2 Eri, HY), 9 stars below *Tianjie* (GS); nine interpreters of dialects exercising their function on a market.

103. *Tianjie* 天節 (90 Tau, HY), 8 stars South of *Fu'er* in *xiu Bi* 畢; a celestial mace; the sceptre of *Shen*.

104. *Jiuliu* 九游 (? Ori), 9 stars South-West of *Yujing* (SS); nine flags.

105. *Junjing* 軍井 (κ Lep, HY), 4 stars South-East of *Yujing* (SS) according to *Jin shu*; the well for soldiers.

106. *Shuifu* 水府 (ν Ori, HY), 4 stars South of *Jing*; the water palace, probably the palace of the River God *He Bo* 河伯.

107. *Sidu* 四瀆 (ϵ Mon, HY), 4 stars South of *xiu Jing*; the four rivers.

    Sometimes these four stars were related to the four rivers of China: *Jiang* 江 (the Yangtze river), *He* 河 (the Yellow river), *Huai* 淮 (the Huai river) and *Si* 泗 (the Si river). See the *Jin shu*.

108. *Queqiu* 闕丘 (δ Mon, HY), 2 stars South of *Nanhe* (SS); two hills on both sides of a gate leading to the palace.

109. *Tiangou* 天狗 (GC10328 Mon, HY), 7 stars North-East of *Lang* 狼 (SS); a celestial dog.

110. *Zhangren* 丈人 (ϵ Col, HY), 2 stars South-West of *Junshi* (SS); husbandman.

111. *Zi* 子 (β Col, HY), 2 stars East of *Zhangren*; son.

[11]The Dunhuang star map of the Tang indicated this 9 star constellation of *Gan Shi* with the name *Tianpu* 天圃, meaning "vegetable garden." This might refer to the garden of *Nanji Laoren* 南极老人, where the Old Man grew his herbs of longevity (*xian chao* 仙草). During the Han both *Tianpu* and *Laoren* (α Car) were visible low above the southern horizon in later evenings of October.

112. *Sun* 孫 (δ Col, HY), 2 stars East of *Zi;*
grandsons.

Zhangren, Zi and *Sun* together represent a farmer's family

113. *Tianshe* 天社 (ν Pup, HY), 6 stars South of *Hu* (SS);
celestial house altar for offerings to the ancestors

114. *Tianji* 天記 (12 Hya, HY), 1 star South of *Waichu* (GS);
a judge to estimate the age of animals.

This judge estimated the age of animals before preparation in *Waichu,* the outer kitchen. It was said that for sacrifice one should not kill young animals. See the *Kaiyuan zhanjing.*

115. *Waichu* 外廚 (2 Hya, HY), 6 stars South of *xiu Liu;*
the outer kitchen in the fields, probably for the occasion of sacrificing to the ancestors. *Tianshe*, *Tianji* and *Waichu* are all in one lunar lodge *xiu Liu*, and belong together. They are probably indications of a Mid-winter sacrifice (on the lunar calendar).

116. *Tianmiao* 天廟 (α Pyx, HY), 14 stars South of *xiu Zhang;*
the celestial ancestor temple.

Jin shu mentions great festivals and ceremonies about *Tianmiao* together with the functioning of *Waichu* in the second half of July (on the solar calendar).

117. *Dong'ou* 東甌 (q Vel, HY), 5 stars South of *xiu Yi*;
the name of a place in the South-Eastern Coastal Area.

Gan Shi used this constellation to indicate territories of barbarians.

118. *Qifu* 器府 (? Vel), 32 stars South of *xiu Zhen;*
the house for musical instruments and vessels. See *Jin shu.*

Musical instruments and vessels were used in all kinds of ceremonies for offering and sacrificing. The 32 stars were in the deep South of the sky during the Han. Later they became virtually invisible because of precession. All later star maps indicated them in wrong positions so as to maintain the name of the constellation. See the Dunhuang star map for example, on which they are 20 degrees too far North..

A bas-relief from the Former Han showing the performance of music, dancing and acrobatics in ancient China. These games with musical accompaniment were called *bai xi*, ''A Hundred Kinds of Games.'' This picture may be related to the constellation *Qifu* of *Gan Shi*.

APPENDIX III

THE CONSTELLATIONS OF WUXIAN SHI

The key to this Appendix is the same as for Appendix II.

1. *Taizun* 太尊 (GC14737 UMa, HY), 1 star North of the middle "step" of *Santai* (SS);
   a honorable and respected old man.

   Taizun is on the same hour-circle as the temple for *Ji* 稷 in the deep South. There were big thanksgiving festivals for Hou Ji 后稷 in *da shu jie* 大暑節 (about 1st August). So *Taizun* was probably identical with Hou Ji.
2. *Sangong* 三公 or *Sanshi* 三師 (30 UMa, HY), 3 stars to the West of the first star of *Beidou* (α UMa);
   three top instructors, not mentioned by Sima Qian.
3. *Dali* 大理 (?), 2 stars inside *Ziwei;*
   High Judge of the Qin court. See also Appendix II, 12.
4. *Nüyu* 女御 (?), 4 stars near *Gouchen* 鉤陳 (SS);
   educated court ladies, teaching the imperial concubines how to behave.
5. *Tianxiang* 天相 (γ Sex, HY), 3 stars North of *xiu Xing*;
   some ministers.
6. *Changyuan* 長垣 (53 Leo, HY), 4 stars West of *Shaowei* (SS), lining up from South to North;
   a strong wall across the path of the Solar God Huangdi, protecting *Taiwei*.

   In Appendix I, sub nr. 45, we have seen that the leading star of *Shaowei* was 53 Leo. From the description by *Shi Shi* we know for certain that *Shaowei* was identical with 53, 52, 51 Leo and 41 LMi. Therefore between *Shaowei* and *Xuanyuan* 軒轅 (SS) there are no stars which can be shaped in the way described by *Wuxian Shi*. Probably *Wuxian Shi*'s "West of *Shaowei*" should be read as "West of *Taiwei*." In that case *Changyuan* should be identical with *Shaowei*. Indeed the *KYZJ* quoted Xi Meng saying " the most southern star of *Changyuan* is *also* six or seven degrees[1] west of the wall of *Taiwei*." That was exactly the position of 53 Leo. By the character *ye* 也 ("also") we know Xi Meng had already identified *Changyuan* with *Shaowei*.

The indication "West of *Shaowei*" by *Wuxian Shi* created confusion for later astronomers. Zhou Cong (c. 1054) of the Song identified *Shaowei* with 60 Leo as the leading star. Then *Shaowei* became the four stars 73,60, 54 Leo and 44 Lmi and *Changyuan* kept its original place, which eliminated contradiction in the description.

7. *Huben* 虎賁 (72 Leo, HY), 1 star South of the lower "step" of *Santai*; brave guards with tiger-masks. They guarded the Court of *Taiwei*.
8. *Junmen* 軍門 (ξ Hya), 2 stars West of *Qingqiu*;
   a gate for the army of the Red General, Chi Yi 赤羿.
9. *Tusikong* 土司空 (*o* Hya), 3 stars South of *Junmen*;
   a master of constructions.
   Chi Yi's main function when he was with Huangdi was to be his Master of Constructions.

   The constellations *Junmen* (WS), *Tusikong* (WS), *Dong'ou* (GS) and *Qingqiu* (GS) immediately South of *Zhen*, *Yi* and *Zhang* reflect the great interest of *Gan Shi* and *Wuxian Shi* in the mythology about the Archer Yi who was named respectively Chi Yi 赤羿, Sheng Yi 神羿 and Hou Yi 后羿.
10. *Yangmen* 陽門 (51 Hya), 2 stars North-East of *Kulou* (SS);
    a gate at the Eastern side of the carriage road *Ping* (SS).
11. *Dunwan* 頓頑 (58 Hya), 2 stars South-East of *Zhewei* (GS);
    an officer to maintain discipline in the army, authorized to arrest people.

[1]The original gave the distance in the length unit *chi* 尺. One *chi* corresponded with one degree.

12. *Congguan* 從官 (χ Lup), 2 stars South of *xiu Fang;* a page or "*aide de camp*", *KYZJ* described it as a medical officer to cure sick and wounded soldiers.

13. *Tianfu* 天輻 (υ Lib, HY), 2 stars West of *xiu Fang*; reserve spokes for carriage wheels.

14. *Jianbi* 鍵閉 (ν Sco, HY), 1 star North-East of *xiu Fang*; a locked passage to the world of darkness, guarded by a royal officer. The officer keeps the key according to *Jin shu.*

15. *Fa* 罰 (χ Sco), 3 stars West of *Dongxian* (SS); a whip to punish vendors in case of fraud or deceit on the market.

16. *Liesi* 列肆 (λ Oph, HY), 2 stars North-West of *Hu* (GS) in *Tianshi*; a street with many settled small shops, possibly a colonnade.

17. *Chesi* 車肆 (20 Oph, HY), 2 stars close to the first "left" star (ζ Oph) of *Tianshi*; "left" when facing southwards from the position of *Dizuo* (SS).
shops to buy and repair wagons.

18. *Bodu* 帛度 (96 Her, HY), 2 stars North-East of *Zong* (SS); calibrated ruler used to measure textile, indicating cloth vendors.

19. *Tusi* 屠肆 (95 Her, HY), 2 stars North of *Bodu;* pedlars of fresh meat, butcher shop.

20. *Xizhong* 奚仲 (κ Cyg, HY), 4 stars North of *Tianjin* (SS); a legendary charioteer using horses instead of bulls.

    Domestication of horses started in Asia in the second millennium BC, about the transition from the Xia to the Shang. At the beginning of the Zhou the Nomads introduced riding on horseback.

21. *Gouxing* 鉤星 or *Tiangou* 天鉤 (β Cep, HY), 9 stars North of *Zhaofu* (GS);
the celestial hook.

    Jin Shu related *Gouxing* with the detection of earthquakes; the nine stars acted as a chain pendulum to signal low frequency movements of the floor[2]

[2]In 131 AD Zhang Heng invented a seismograph, which was based on "a central bar capable of lateral displacement along tracks in eight radial directions, arranged to operate a closing and opening mechanism, hidden in a closed vessel (*Hou han shu*)."

22. *Tianfu* 天桴 (θ Aql, HY), 4 stars near *Hegu*;
the celestial drumstick.

23. *Tianyue* 天籥 (4 Sgr, HY), 7 stars West of the second star of *Dou*;
the gate in a narrow pass.

Tianyue is in exact opposition with *Tianguan* (SS), both along the celestial path of Sun and Moon.

24. *Tianyuan* 天淵 (α Sgr), 10 stars South-East of *Bie* 鱉 (SS);
a deep place in the Celestial River.

25 – 36 *Shi'erguo* 十二國 (?), here are the twelve governors of the 12 provinces renewing their fiefs. These provinces were

Zhao 趙, Wei 魏, Han 韓, Qi 齊, Zheng 鄭, Yue 越,
Zhou 周, Qin 秦, Dai 代, Jin 晉, Chu 楚 and Yan 燕.

Zhao, Wei and Han were separated from the large state Jin in 403 BC. Zhou was formerly the home-state of the Zhou dynasty. Now it ranked among the states which had become provinces of the Han empire. Dai was formerly a very small state among the Warring States. Now it was as important as those big states simply because Han Wu Di was once the governor of the province of Dai. These facts provide valuable information about the time of introduction of the term *Shi'erguo* which certainly was after Sima Qian's *Tianguan shu. Shi'erguo* was mentioned for the first time in the *Jin shu*.

37 *Liyu* 離瑜 (?), 3 stars East of Dai 代, one of the *Shi'erguo*;
jade ornaments to wear on clothes; decorations.

38 *Tianleicheng* 天壘城 (?), 13 stars South of *Ku* (GS) and *Qi* (GS);
celestial castle with earthern walls.

39 *Xuliang* 虛梁 (κ Aqr, HY), 4 stars South of *xiu Wei* 危;
a house without pillars, dome-shaped like in a mausoleum. See the *KYZJ*.

Eight dragon heads distributed around the upper rim of the vessel held eight bronze balls in their mouths. A seismic tremor induced the central pendulum to move in a certain lateral direction, pushing a ball out of a specific dragon's mouth by means of a radial bar. The ball fell with a loud bang into the open mouth of an underlying bronze toad, warning the attendants about a disaster and the direction in which it happened. The same principle was used by Heron of Alexandria about 62 AD. See Needham [1959].

40 *Tianqian* 天錢 (?), 10 stars North-West of *Beiluo shimen* (SS); celestial paper money to be burned in *Baijiu* (SS); at mid-winter spirits and ghosts had to be appeased.

41 *Tiangang* 天綱 (δ PsA), 1 star South-West of *Beiluo shimen;* facilities to make tents for the people living in the winter-camp. See the *KYZJ*.

42 *Fuyue* 鈇鉞 (β Scl, HY), 3 stars North-West of *Bakui* (GS); sickle-shaped axes for the soldiers.

43 *Tianjiu* 天廄 (22 And, HY), 10 stars North of constellation *xiu Bi* 壁, close to *Wangliang* (SS);
the celestial horse stable along *Gedao* (SS).

Jin shu mentions *yi ting* 驛亭 (post offices) to provide messengers with fresh horses. A messenger of the King was said to "beat" the clock in speed. This system for the fast transmission of messages dates from the Qin.

44 *Tianyin* 天陰 (δ Ari, HY), 5 stars West of *xiu Bi* 畢;
the celestial *yin* force.

The meaning of *Tianyin* is obscure. It probably means the dark northern side of a hill. *Tian'e* (GS) is nearby and also means the slope of a hill.

BIBLIOGRAPHY

The Chinese sky was such a large cultural complex that the study of it actually concerns every aspect of ancient culture. Thus a bibliography could be huge. That is not our purpose here. The bibliography that follows is not exhaustive, but includes only those works most relevant to the subject treated here. We want to emphasize that our work is primarily based on available Chinese sources, thus very probably we have missed some recent valuable publications on the relevant subjects.

Chinese primary sources are arranged alphabetically by title in *pinyin* spelling. Each item includes information in the sequence of: (i) title in *pinyin* spelling and in Chinese characters; (ii) translation of the title, with short explanations of relevance in parentheses where required; (iii) the dynasty and the year when the source came into existence, (in some cases they are not certainly known); (iv) authors or compilers of the source; (v) important western language translation to which we have referred.

Secondary Chinese & Japanese and Western sources are alphabetized in the conventional way, by authors. Of course the *pinyin* spelling is applied to Chinese authors.

A. Chinese Primary Sources

Bu tian ge 步天歌
Pacing the Heavens.
Sui, c. +590.
Wang Ximing 王希明.

Chidao nanbei liangzong xingtu 赤道南北兩總星圖
General Star Atlas of Southern and Northern Hemispheres divided by the Equator.

Late Ming and Early Qing (affiliated to the *Chongzhen lishu,* Astronomical Treatise of the Chongzhen era) +1634.
Tang Ruowang 湯若望 (Johan Adam Schall von Bell).
a planisphere preserved in the Vatican library.

Chongzhen lishu 崇禎曆書
The Astronomical Encyclopedia of the Chongzhen reign period.
Ming, +1635.
Xu Guangqi 徐光啓 & Tang Ruowang (Johann Adam Schall von Bell).
Later was edited and presented to the Qing court by Tang Ruowang as *Xiyang xinfa suanshu.*

Chouren zhuan 疇人傳
Biographies of Mathematicians and Astronomers.
Qing, +1790.
Yuan Yuan 阮元.

Chu ci 楚辭
Elegies of Chu (State).
Zhou (Zhan Guo period), c.-300.
Qu Yuan 屈原.

Chunqiu fanlu 春秋繁露
Luxurant Dew of the Spring and Autumn Annals.
Former Han, -145.
Dong Zhongshu 董仲舒.

Dunhuang xingtu 敦煌星圖
Star maps in Dunhuang collection from the Qianfugong 千佛洞 grottoes in Gansu 甘肅 province, discovered by M.A. Stein in 1907; in British Museum sub nr. 3326.
Tang, c. +8th century.

Er ya 爾雅
Literary Expositor [dictionary].
Former Han.
Compilers unknown.
Commented on c. +300 by Guo Pu 郭璞.

Fa yan 法言
Admonitory Sayings [in imitation of *Lun Yu*].
Xin, +5.
Yang Xiong 揚雄.

Fengshu tongyi 風俗通義
On Popular Traditions and Customs.
Later Han, +175.
Ying Shao 應劭.

Ganshi xingjing 甘石星經
or *Tongzhan daxiangli xingjing* 通占大象曆星經
Star Manual ofthe Masters Gan (De) and Shi (Shen).
Northern Zhou, c. +579.
A compilation of star-catalogues from the Han or even earlier, attributed to Gan De and Shi Shen.
Ma Xian 馬顯.

Guji luli kao 古今律曆考
Investigation of (Chinese) Calendars, Old and New.
Ming, c. +1600.
Xing Yunlu 邢雲路.

Guo yu 國語
Sayings of the States.
A collection of historical conversations arranged under states, and covering about the Chunqiu period (771-481 BC).
Compilers unknown.

Han shu 漢書
History of the Former Han (-206 to +24).
Later Han, c. +100.
Ban Gu 班固.

Hou han shu 後漢書
History of the Later Han dynasty (+25 to +220).
Liang Song, c. +450.
Fan Ye 范曄.
Hou han shu lulizhi 後漢書 • 律曆志
Treatises on Harmonics and Calendars; actually they were chapters of the *Xu han shu* (The Continued History of the Han) written by Sima Biao 司馬彪 (c. 240 - c. 306) and they were incorporated into the *Hou han shu* by Liu Zhao 劉昭 (fl. 502 - 520).

Huainan zi tianwenxun 淮南子 • 天文訓
The Astronomical Chapter of the Book of the Prince of Huai Nan.
Former Han, c. -120.

Written by a group of scholars gathered around Liu An 劉安, prince of Huainan.

Huangdi neijing suwen 黄帝内經素問
Pure Questions of the Yellow Emperor, the Canon of Internal Medicine.
Qin, Former Han or Later Han.
Writers unknown.

Huntianyi zhu 渾天儀注
Notes on the Armillary Sphere.
Later Han, c. +117.
Zhang Heng
(Fragment existing in commentaries on the *Hou han shu lulizhi* by Liu Zhao.)

Jin shu 晉書
History of the Jin dynasty.
Tang, +635.
Fang Xuanling.
Jin shu tianwenzhi 晉書 • 天文志
The astronomical chapter of the *Jin shu*.
Tang, c. +628.
Li Chunfeng 李淳風.
Tr. Ho Peng-Yoke [1965].

Jing tian gai 經天該
Rhymed Catalogue of Stars.
Ming, c. +1612.
Li Madou 利瑪竇(Matteo Ricci) & Li Zhizao 李之藻.

Jiu tang shu 舊唐書
Old History of the Tang dynasty.
Five Dynasties, +945.
Liu Xu 劉昫.

Kaiyuan zhanjing 開元占經
Treatise on Astrology of the Kaiyuan reign period.
Tang, +729.
Qutan Xida 瞿曇悉達.

Li ji 禮記
Records of Rituals.
Former Han, c. -70; containing material from the Zhou.

Attrib. to Dai De 戴德.

Lingtai miyuan 靈臺秘苑

The Secret Garden of the Observatory (including a star catalogue from the Song).

Bei Zhou, c. +580.

Yu Jicai 庾季才, supplemented by Wang Anli 王安禮, c. +1052.

Lingtai yixiang zhi 靈臺儀象志

On the Astronomical Instruments in the Imperial Observatory.

Qing, +1674.

Nan Huairen 南懷仁 (Ferdinand Verbiest).

Ling xian 靈憲

Spiritual Constitution of the Universe.

Later Han, c. +118.

Zhang Heng 張衡.

Existing in fragment.

Lüshi chunqiu 呂氏春秋

Master Lü's Spring and Autumn Annals.

Zhou (Qin State), c. -239.

Lü Buwei 呂不韋.

Lun heng 論衡

Discourse Weighed in Balance.

Later Han, c. +83.

Wang Chong 王充.

Tr. Forke [1907].

Lun yu 論語

Confucius Analects.

A collection of sayings by Confucius (551 - 479 BC) and some of his disciples recorded by disciples of the Confucian School.

Tr. Legge [1861].

Ming shi tianwenzhi 明史・天文志

Treatise on Astronomy of the *Ming shi* (History of the Ming Dynasty).

Qing, +1723.

Mei Juecheng 梅瑴成.

Mingyi tianwen shu 明譯天文書

Astronomical Books Translated (by Imperial order of the) Ming.

Ming, +1382.

Hai Da'er 海達兒.

Mutianzi zhuan 穆天子傳

Travels of Emperor Mu to the West.

Zhou, before -245.

Writer unknown.

Qizheng tuibu 七政推步

On the Motions of Seven Luminaries [planetary ephemerides prepared according to the methods of the Muslim astronomers in China].

Ming, +1477.

Bei Lin 貝琳.

Qian han shu; see *Han shu.*

She-tou-jie taizi ershiba xiu jing 舍頭諫太子二十八宿經

Classics on the Twenty Eight Lunar Lodges by Prince She-tou-jie.

Translated from Sanskrit.

Jin, c. +300.

Translater unknown.

Shi ji 史記

Historical Records.

Former Han, ca. -90

Sima Qian 司馬遷.

Tr. Chavannes [1898]

Shijia zhai yangxin lu 十駕齋養新錄

Notes about New Things Learned from the Old in the Study Shijia.

Qing, c. +1700.

Qian Daxin 錢大昕.

Sanfu huangtu 三輔黃圖

Description of the Three Districts of the Capital (Chang'an).

Jin, late +3rd century.

Attrib. to Miao Changyan 苗昌言.

Shanhai jing 山海經

Classic of Mountains and Rivers.

Zhou and Former Han.

Writers unknown.

Shi jing 詩經

Book of Odes.

Zhou, -9th to -5th century.

Writers unknown.
Tr. Legge [1871].

Shuowen jiezi 說文解字
Analytical Dictionary of Characters.
Later Han, +121.
Xu Shen 許愼.

Suzhou tianwen tu 蘇州天文圖
Suzhou Astronomical Chart; a stone planisphere from the Wen Miao temple based on maps of Huang Shang (黄裳, +1193).
Tr. Rufus [1945].

Shu jing 書經
Book of Documents.
Zhou, -5th century, containing material from very ancient times; text stabilised in Qin or Former Han.
Writers unknown.
Tr. Legge [1865].

Song shi 宋史
History of the Song Dynasty.
Yuan, c. +1345.
Tuo-Tuo 脱脱 & Ouyang Xuan 歐陽玄.

Song shu 宋書
History of the (Liu) Song Dynasty (+420 to +478).
Southern Qi, +500.
Shen Yue 沈約.

Sui shu tianwenzhi 隋書 • 天文志
The astronomical chapter of *Sui shu*
Tang, c. +628.
Li Chunfeng 李淳風.

Tiandi ruixiang zhi 天地瑞祥志
Record of Auspicious Phenomena in Heaven and on Earth.
Tang, c. +666.
Sa Shou-Zeng 薩守眞.

Tiangong kaiwu 天工開物
The Use of the Works of Nature.
Ming, +1637.
Song Yingxing 宋應星.

Tianwen dacheng guankui jiyao 天文大成管窺輯要
Essentials of Astronomy.
Qing, incorporating Song and Yuan material, +1653.
Huang Ding 黄鼎.
Tianwen huichao 天文彙鈔
A Collection of Astronomical Texts.
Ming, with a star-catalogue from the early Ming.
Copied by an unknown Ming-scholar.
Preserved in Beijing Library.
Tianwen yaolu 天文要錄
Record of the Most Important Astronomical Matters.
Tang, c. +664.
Li Feng 李鳳.
Tianyuan lili quanshu 天元曆理全書
Complete Treatise of the Tianyuan Calendar.
Qing, +1682.
Xu Fa 徐發.
Tong dian 通典
Comprehensive Institutes (reservoir of source material on political and social history).
Tang, +812.
Tu You 杜佑.
Wen xuan 文選
General Anthology of Prose and Verse.
Liang, +530.
Ed. Xiao Tong 蕭統 (prince of the Liang).
Wuxing dayi 五行大義
Fundamental Principles of the Five Elements.
Sui, c. +600.
Xiao Jie 蕭吉.
Xinfa suanshu 西洋新法算書
Treatise on Mathematics (Astronomy and Calendrical Science) according to the New Methods.
Qing, +1628, revised in +1644.
Tang Ruowang 湯若望 and Xu Guangqi 徐光啓.
Xia xiao zheng 夏小正

Small Calendar of the Xia dynasty
Zhou, between -7th and -4th centuries, containing astronomical information from the Xia dynasty.
Writers unknown. Tr. Soothill [1951].

Xin tang shu 新唐書
New History of the Tang dynasty.
Song, +1061.
Ouyang Xiu 歐陽修 & Song Qi 宋祁.

Xin yixiang fayao 新儀象法要
New Design for Astronomical Instruments [with a set of star-maps].
Song, +1094.
Su Song 蘇頌.

Xu han shu lulizhi 續漢書 • 律曆志
See *Hou han shu*.

Xun zi 荀子
Book of Master Xun.
Zhou; Zhanguo period, c. -240.
Xun Qing 荀卿.

Yi jing 易經
Book of Changes.
Zhou with Former Han additions.
Compilers unknown.

Yixiang kaocheng 儀象考成
Imperial Astronomical Instruments. [containing a star catalogue.]
Qing, +1744.
Dai Jinxian 戴進賢 (Ignatius Kögler) and others in the Imperial Astronomical Bureau.

Yue ling 月令
Monthly Ordinances.
Qin, Former Han, c. -230; containing astronomical information from early Zhou.
Writers unknown.
Existing as chapters in the *Lüshi chunqiu* and the *Li ji*.

Zhanguo ce 戰國策
Records of the Warring States.
Qin.

Writer unknown.
Zhongxi jingxing tongyi kao 中西經星同異考
Similarities and Differences between Chinese and Western Star-Names.
Qing, +1723.
Mei Wending 梅文鼎.
Zhoubi suanjing 周髀算經
Arithmetical Classic of the Circle and the Gnomon
Former Han.
Writers unknown.
Zhou li 周禮
Rituals of the Zhou dynasty [descriptions of the Zhou officialdom].
Former Han, containing some material from late Zhou.
Compilers unknown.
Tr. E. Biot [1851].
Zuo zhuan 左傳
Master Zuo's Tales about the *Chun qiu* (The Spring and Autumn Annals).
Chou; Zhanguo period, compiled between -430 and -250, but probably modified by Confucian scholars of the Han, especially 劉歆.
Attrib. to Zuo Jiuming 左丘明.

B. Chinese and Japanese Secondary Sources

Bo Shuren 薄樹人
"Zhongguo Gudai de hengxing guance" 中國古代的恆星觀測
On the observation of fixed stars in ancient China,
Kexue shi jikan 科學史集刊
Studies on History of Science,
1960, No.3, 35-52.
Bo Shuren 薄樹人
"Sima Qian – wuguo gudai weida de tianwenxue jia"
司馬遷－我國古代偉大的天文學家
The great Chinese astronomer Sima Qian,
Zhiran zhazhi 自然雜志 (Nature),
4 (1980), No.9, pp.685-688.
Bo Shuren 薄樹人 (chief editor)

Zhongguo tianwenxue shi 中國天文學史
A History of Astronomy of China.
1981, Science Press, Beijing.

Chen Jiujing 陳久金, Lu Yang 盧央, Liu Yaorang 劉堯讓
Yizhu tianwenxue shi 彝族天文學史
A History of Astronomy of the Yi-people.
1984, Yunnan Press.

Chen Meidong 陳美東
"Richanbiao zhi yanjiu" 日躔表之研究
Studies on the tables on solar motion (in ancient China).
Zhiran kexueshi yanjiu 自然科學史研究
Studies in the History of Natural Sciences,
3 (1984), No.4, 330-340.

Chen Meidong 陳美東
"Chen Zhuo xingguan de lishi shanbian" 陳卓星官的歷史嬗變
The evolution of Chen Zhuo's constellations, *Kejishi wenji* 科技史文集, Collected Studies on History of Science and Technology,
16 (1992), 77-91.

Chen Ying 陳鷹
"Tianwen huichao xingbiao he Guo Shoujing de hengxing guance gongzuo"
天文彙鈔星表和郭守敬的恆星觀測工作
The *Tianwen huichao* star catalogue and Guo Shoujing's observations,
Zhiran kexueshi yanjiu 自然科學史研究
Studies in the History of Natural Sciences,
5 (1986), No.4, 331-340.

Chen Zungui 陳遵媯
Zhongguo tianwenxue shi 中國天文學史
History of Astronomy of China, 4 volumes.
1980, 1982, 1984, 1989, Shanghai.

Ding Shan 丁山
Zhongguo gudai shenghua 中國古代神話
Myths of Ancient China.
1961, Shanghai.

Du Shengyun 杜升雲
"Suzhou shike tianwentu hengxing weizhi de yanjiu"
蘇州石刻天文圖恆星位置的研究

On the position of the stars on the Suzhou astronomical chart),
Beijing shifan daxue xuebao 北京師範大學學報
Journal of Beijing Normal University,
1982, No.2, p.81.

Feng Shi 馮時
"Zhongguo zhaoqi xingxiangtu yanjiu" 中國早期星象圖研究
A study of early star charts in China,
Zhiran kexueshi yanjiu 自然科學史研究
Studies in the History of Natural Science,
9 (1990), No.2.

Gao Lu 高魯
Xingxiang tongjian 星象統箋
Evolution of Chinese constellations.
National Research Institute for Astronomy, monograph No.2. 1933.

Gu Jigang 顧頡剛
Qin Han de fangshi he rushen 秦漢的方士和儒生
Magicians and Confucian Scholars during the Qin and the Han.
1955, Junzhong Press, Shanghai.

Hashimoto Keizō 橋本敬造
赤道南北兩總星圖と恆星屏幛
"General Star Atlas of Southern and Northern Hemispheres divided by the Equator" and "Planispheres of the Fixed Stars"
新發現中國科學史資料研究・論考篇
Studies on recently discovered material about History of Science, Section of Historical Research.
1985, Kyoto.

Huang Yilong 黄一農
"Jixing yu gudu kao," 极星與古度考
Tsinghua Journal of Chinese Studies,
22 (1992), 93-117.

Iijima Tadao 飯島忠夫
"Shina no jōdai ni okeru Girisha bunka no eikyō to jukyō keiten no kansei," 支那の上代における希臘文化の影響と儒教經典の完成
Greek influence on the ancient civilization of China and the compilation of the confucian classics with reference to calendrical science and astronomy, *Tōyō gakuhō* 東洋學報

Reports of Oriental Society of Tokyo,
11 (1921), No.1, pp.183 ff.
Iijima Tadao 飯島忠夫
"Shu jing, Shi jing zhi tianwen lifa" 書經詩經天文曆法
Astronomy and calendrical science in the *Shu jing* and the *Shi jing*,
Ke xue 科學
Science, 13 (1928).
Jiang Xiaoyuan 江曉原
Tianxue zhenyuan 天學眞原
Prime Themes of (Ancient) Learning of Heaven.
1991, Liaoling, China.
Li Yong 李勇
"Dui zhongguo gudai hengxing fenye he feye shipan de yanjiu" 對中國古代恆星分野和分野式盤的研究
On the *fen ye* theory of constellations and the divination board,
Zhiran kexueshi yanjiu 自然科學史研究
Studies in the History of Natural Science,
11 (1992), No.1, 22-31.
Liu Caonan 劉操南
"Ershiba xiu sheming" 二十八宿釋名
On the meaning of the names of the 28 *Xiu*,
Shehui kexue zhanxian 社會科學戰線
Frontier in Social Science,
1979, No.1, 153-162.
Liu Jinyi 劉金沂 and Wang Jianmin 王健民
"Chen Zhuo he Gan Shi Wu sanjia xingguan" 陳卓和甘石巫三家星官, Chen Zhuo and the constellations of Gan, Shi, and Wu,
Kejishi wenji 科技史文集 (Collected papers on History of Science),
6 (1979), 32-44.
Liu Chaoyang 劉朝陽
"Cong tianwen lifa tuice Yaodian zhi biancheng niandai"
從天文曆法推測堯典之編成年代
On the time of compilement of the Canon of Yao by means of astronomical considerations,
Yanjing xuebao 燕京學報
Journal of Yanjing University,

7 (1930).
Liu Wendian 劉文典
Huainan honglie jijie 淮南鴻烈集解
Collected Commentaries on the Book *Huainan zi.*
1923, Shanghai.
Lu Yang 盧央 et al
"Ming chidao liangzong xingtu jianjie" 明赤道兩總星圖簡解
A brief introduction of the General Star Atlas of Southern and Northern Hemispheres divided by the Equator,
Zhongguo gudai tianwen wenwu lunji 中國古代天文文物論集
Collected Papers on Astronomical Relics in China.
1989, Wem Wu Publishing House, Beijing.
Lü Zifang 呂子方
Lü Zifang kexueshi lunwen ji 呂子方科學史論文集
Collection of Lü Zhifang's Articles on the History of Science.
1984, 2 volumes, Si Chuan.
Ma Shichang
Dunhuang xingtu de niandai 敦煌星圖的年代
On the date of the Dunhuang star-maps,
Zhongguo gudai tianwen wenwu lunji 中國古代天文文物論集
Collected Papers on Astronomical Relics in China.
1989, Wen Wu Publishing House, Beijing.
Noda Chūryō 能田忠亮
"Gyōten ni mietaru temmon" 堯典に見えたる天文
Astronomy in the Canon of Yao,
Tōhō gakuhō 東方學報 (Kyoto),
8 (1937), 118.
Noda Chūryō 能田忠亮
"Kashōsei seishōren" 夏小正星象論
On the astronomy in the *Xia xiao zheng,*
Tōhō gakuhō 東方學報 (Kyoto),
12 (1941), 209.
Noda Chūryō
"Kan-Seki Seikyō kō" 甘石星經考
A study of the star manual of Gan De and Shi Shen,
Tōhō gakuhō 東方學報,

1 (1931).
Osaki Shoji 大畸正次
中國星座の歷史
History of the Chinese Constellations.
1988, Kyoto.
Pan Nai 潘鼐
Zhongguo hengxing guance shi 中國恆星觀測史
History of Observation of Stars in China.
1989, Scholar Press, Shanghai.
Pang Pu 龐樸
"Huoli santan" 火曆三探
Discussions on ancient calendars based on the observation of the star Huo, *Wen shi zhe* 文史哲
Literature, History and Philosophy,
1988, No.3.
Pang Pu 龐樸
"Houli gouchen" 火曆鉤陳
The discovery of the Fire Calendar,
Zhongguo wenhua 中國文化
Chinese Culture,
1 (1989), 1-35.
Qian Baocong 錢寶琮
"Ganshi xingjing yuanliu kao" 甘石星經源流考
On the origin and transition of Astrological Texts of Gan De and Shi Shen, *Guoli Zhejiang daxue jikan* 國立浙江大學季刊
Quarterly publication of the State University of Zhejiang,
6 (1937).
Qian Baocong 錢寶琮
"lun ershiba xiu zhi laili" 論二十八宿之來歷
On the origin of the twenty eight lunar lodges,
Sixiang yu shidai 思想與時代
Ideas and Times,
43 (1947), No.3.
Qian Baocong 錢寶琮
"Gaitian shuo yuanliu kao" 蓋天說源流考
On the origin of the cosmology of Kai Tian (Covering Sky),

Kexueshi jikan 科學史集刊
1958, No.1, Science Press, Beijing.
Shinjō Shinzō 新城新藏
Dongyang tianwenxue shi yanjiu 東洋天文學史研究
Researches on the History of Astronomy in the Far East.
1928, Tokyo; 1933, Chinese translation by Chen Zungui, Shanghai.
Sun Dexuan 孫德萱 et al
"Henan Puyang mu waque baokao" 河南濮陽墓挖掘報告
Reports on the excavation of the Puyang graveyard,
Wen wu 文物
Relics,
1984, No.1.
Tan Qixiang 譚其驤
Zhongguo lishi ditu ji 中國歷史地圖
Historical Geological Atlas of China.
1982, Beijing.
Ueda Yuzuru 上田穰
Seki-shi Seikyō no Kenkyū 石氏星經の研究
Investigations on the Star Manual of Shi Shen.
1929, Tokyo.
Wang Jianmin 王健民, et al.
"Zenghouyi mu chutude ershibaxiu qinglong baihu tuxiang"
曾侯乙墓出土的二十八宿青龍白虎圖象
The twenty eight Lunar Lodges and the images of the Green Dragon and the White Tiger discovered in the tomb of Zeng Hou Yi,
Wen wu 文物 (Relics),
7 (1979), pp. 40-45.
Wen Yiduo 聞一多
Wen Yiduo chuanji 聞一多全集
Complete Work of Wen Yiduo.
1948, Shanghai. Compiled by Zhu Zhiqing 朱自清.
Wu Qichang 吳其昌
"Han yiqian hengxing faxian cidi kao" 漢以前恆星發現次第考
A study of the identification of the fixed stars known before the Han,
Zhenli Zhazhi 眞理雜志
1 (1943), No.3, 273..

Xi Zezong 席澤宗
"Gu xinxing biao" 古新星表
A new catalogue of ancient novae,
Acta Astronomica Sinica,
3 (1955), No.2, p.183.
Xi Zezong 席澤宗
"Suzhou shike xingtu" 蘇州石刻星圖
The Suzhou astronomical chart engraved on stone,
Wenwu chankao zhiliao 文物参考資料
Reference Material for the study of Cultural Relics,
1958, No.7.
Xi Zezong 席澤宗
"Dunhuang xingtu" 敦煌星圖
A star map from Dunhuang,
Wenwu 文物, Cultural Relics,
1966, No.3: 27-38.
Xia Nai 夏鼐
"Cong Xuanhua Lia0mu de xingtu lun ershiba xiu he huangdao shier gong," 從宣化遼墓的星圖論二十八宿和黄道十二宮
On the twenty eight lunar lodges and the twelve zodiacal signs: based on the star map from a Liao-tomb in Xuanhua,
Kaogu xuebao 考古學報
Acta Archaeologica Sinica,
1976, No.2, pp.35-58.
Xu Fuguan 徐復觀
Lianghan sixiang shi 兩漢思想史
History of Ideology during the Former and the Later Han.
Vol.2, 1975, The Chinese University of Hong Kong.
Yabuuchi Kiyoshi 藪内清
"Sō dai no seishuku" 宋代の星宿
The constellations in the Song Dynasty,
Tōhō gakuhō 東方學報,
1 (1936).
Yabuuchi Kiyoshi 藪内清
"Tō Kaigen Seikyō chū no seikyō" 唐開元占經中之星經 The star catalogue in the *Kaiyuan zhanjing*,

Tōhō gakuhō 東方學報 (Kyoto),
8 (1937), pp.56-74.
Yabuuchi Kiyoshi 藪内清
"Kandai ni okeru kansoku gijutsu to sekishi seikei no seiritsu"
漢代における觀測技術と石氏星經の成立
Observational basis for the formation of the star manual of Shi Shi,
Tōhō gakuhō 東方學報,
30 (1959).
Yabuuchi Kiyoshi 藪内清
Chūgoku no tenmon rekihō 中國の天文暦法
Chinese Astronomy and Calendrical Science.
1969, Tokyo.
Yao Zhongyi 饒宗頤
Yunmeng qinjian rishu yanjiu 雲夢秦簡日書研究
A Study of the Bamboo Book from the Qin discovered in Yunmeng.
1981, Hong Kong
Yasui Kozan 安居香山 and Nakamura Shohachi 中村幛八
緯書基礎の研究
A Fundamental Study on the *Wei-shu* ("woof" books).
1976, Enlightenment Press, Kyoto.
Yi Shitong 伊世同
Zhongxi hengxing duizhao tubiao 中西恆星對照圖表
Correlative Star Catalogue and Atlas of China and the West.
1981, Science Press, Beijing.
Zhao Yuanren 趙元任
"Zhongxi xingming tukao" 中西星名圖考
A comparative study on Chinese and Western star-names.
Ke xue 科學
3 (1917), No.3.
Zhu Kezhen 竺可楨,
"Lun yi shuica ding Shangshu yaodian sizhongxing zhi niandai"
論以歲差定尚書堯典四仲星年代 On the epoch of the four cardinal points recorded in the Canon of Yao,
Ke xue 科學
11 (1926), No.12.
Zhu Wenxin 朱文鑫

Shiji tianguan shu hengxing tukao 史記天官書恆星圖考
A Study of the Constellations Recorded in the astronomical chapter of *Shi ji*.
1927; reprinted 1934, Shanghai.

C. Western Language Sources

Allen, Richard H. *Star Names, Lore and Meaning*; New York, 1963 (reprint of the original edition published by E. Stechert; New York, 1899).

Arakawa, H., *World Survey of Climatology, Vol. 8: Climates of North and East Asia*; Elsevier, Amsterdam, 1969.

The Astronomical Almanac of Washington; Washington, 1995, B16-18.

Bentley, J. *Hindu Astronomy: A Historical Review*; London, 1825.

Biot, J. B., *Le Tcheou-Li ou Rites des Tcheou,* 3 volumes; Paris, 1851.

Biot, J. B., *Étude sur l'Astronomie Indienne et sur l'Astronomie Chinoise;* Paris, 1862.

Blondeau, R. A. *Ferdinand Verbiest*; Ed. Lannoo, Tielt and Bussum, Belgium 1983.

Boss, B. *General Catalogue of 33342 Stars for the Epoch 1950*; Publications of the Carnegie Institute of Washington, nr. 468, 5 volumes, 1936-'37.

Burgess, E. *Translation of the* Surya Siddhanta; Calcutta, 1935.

Chang, C. S. "Military aspects of Han Wu Di's northern campaigns," *Harvard Journal of Asiatic Studies*, 26 (1966), 148.

Chatley, H. "Ancient Chinese astronomy (calendrical)," *Occasional Notes of the Royal Astronomical Society*, 1 (1939), no.5.

Chatley, H. "The Lunar Mansions in Egypt," *ISIS,* 3 (1940), 394-397.

Chavannes, E. *Les Memoires Historiques de Se-Ma Ts'ien,* tome III; Ernest Leroux, Paris, 1898 (repr. E.J. Brill, Leiden 1967).

Chavannes, E. "Carte astronomique d'un empereur de Chine en l'an 1193," *Memoires Concernant L'Asie Orientale,* vol. 1; Paris, 1913.

Chen, Cheng-Yih; Xi, Zezong, "The *Yao Dian* and the origins of astronomy in China," in Clive L.N. Ruggles, Nicholas J. Saunders (eds.) *Astronomies and Cultures.* Niwot: Univ. Press of Colorado, 1993, 32-66.

Cleaves, F. W., *The Secret History of the Mongals*; Harvard Univ. Press, Cambridge Mass. 1982.

Creel, H. G., *Chinese Thought*; London, 1962.

Cullen, C., "Motivations for scientific change in ancient China: Emperor Wu and the grand inception astronomical reforms of 104 BC," *Journal for the History of Astronomy,* xxiv (1993), pp.185-203.

Diény, J. P., *Le Symbolisme du Dragon*; College de France, Paris, 1987.

Ecsedy, I., Barlai, K., Dvorak, R. and Schult, R., "Antares Year in Ancient China," *World Archeoastronomy*; Cambridge Univ. Press, Cambridge, 1987.

Edkins, J., "Babylonian origin of Chinese astronomy and astrology," *China Review,*, 14 (1885), p.90.

Edkins, J., "Star names among the ancient Chinese," *China Review,* 39 (1888), p.309.

Flamsteed, J., *Historia Coelestis Britannica*, 1712; and *Atlas Coelestic (Catalogue),* 1729; published by Greenwich Observatory.

Forke, A., *Lun Heng, Philosophical Essays of Wang Chong;* Shanghai, 1907.

Fraser, J. T., *Of Time, Passion and Knowledge*; Edit. G. Braziller, New York, 1975.

Gain, D.B., *Aratus Phaenomena*; Athlone Press, London, 1976.

Gaubil, A., *Traité de l'Astronomie Chinoise*; Paris, 1732.

de Groot, J. J. L., *The Religious System of China*, 6 volumes; Brill, Leiden, between 1892 and 1910.

Gurshtein, A. A., "On the origin of the zodiacal constellations," *Vistas in Astronomy,* 36 (1993), pp.171-190.

Hartner, W., "The role of observations in ancient and medieval astronomy," *Journal of History of Astronomy,* 8 (1977): 1-11.

Hartner, W., "The astronomical instruments of Cha-Ma-Lu-Ting, their identification, and their relations to the instruments of the Observatory of Maragha," *Isis,* 41 (1950), 184.

Hartner, W., "The obliquity of the ecliptic according to the *Hou han shu* and Ptolemy. Pub. in silver Jubilee volume of Zinbum Kagaku Kenkyu-Syo, Kyoto University, 1954, p. 177.

Hashimoto, Keizo, *Galileo's Invention and Kepler's Optical Astronomy at the Compilation of the Ch'ung-Chen Li-Shu Astronomical Encyclopaedia: 1629-35;* Kansai University, 1986.

Hashimoto, Keizo, *Hsü Kuang-Ch'i and Astronomical Reform,* Kansai University Press, 1988.

Heidel, A., *The Babylonian Genesis*; Univ. of Chicago Press, 1942, 1963.

Heissig, W., *The Religion of Mongolia*; Routledge and Kegan, Lodon, 1980.

Hoffleit, D.; Jaschek, C., *The Bright Star Catalogue*; Yale University Observatory, New Haven, Connecticut, U.S.A. 1982.

Ho Peng Yoke, *Astronomical Chapter of Jin Shu*; Mouton and Co. Paris-The Hague, 1966.

Hulsewe, A. F. P., *Remnants of Han Law, Sinica Leidensia,* Editor Brill, Leiden, 1955.

Kalinowski, M., *Cosmologie et divination dans la Chine ancienne: le compendium des cinq agents (Wuxing dayi, VI siecle). Traduit et annoté.* (Publications de L'École Française d'Extrême-Orient, 166.) Paris, 1991.

Karlgren, B. "Legends and cults in ancient China," *Bull. Mus. Far Eastern Antiquities*, 18 (1946): 200.

Karlgren, B., *Shijing, the Book of Odes;* Mus. Far Eastern Antiquities, Stockholm, 1950.

Kingsmill, T. W., "The Chinese calendar, its origin, history and connections," *Journal of the China Branch of the Royal Asiatic Society,* 32 (1897), pp. 1-35.

Kingsmill, T. W., "Comparative tables of ancient Chinese lunar asterisms," *Journal of the China Branch of the Royal Asiatic Society,* 26 (1892), pp.44-47.

Kistemaker, J. and Yang Zhengzong, "A new approach to traditional Chinese astronomy," *Proceedings Int. Astron. Union, Symp. 133, Mapping the Sky*, Paris 1987.

Kramers, R. P., "The Confucian school", in *Cambridge History of China,* vol.1; Cambridge, U.K., 1987.

Kunitzsch, P., *Untersuchungen zur Sternnomenklatur der Araber;* Wiesbaden, 1961.

Kunitzsch, P., "Al-Manazil," *Encyclopedia of Islam,* Vol. VI, Leiden, 1991, pp.374-376.

Lai, T. C. and Mok, R.; *The Jade Flute, a Story of Chinese Music;* Schocken books, New York, 1985.

Legge, J., *The Chinese Classics, etc.;* vol.1: *Confucian Analects, The Great Learning, and the Doctrine of the Mean,* Hongkong, 1861; vol.3: *The `Shoo King' [Shu Ching],* Hongkong, 1865; vol.4: *`Shih Ching'; the Book of Poetry,* Hongkong-London, 1871.

Loewe, M., *Chinese Ideas of Life and Death;* Allen and Unwinn, London, 1982.

Mackenzie, D. A., *Myths and Legends of China and Japan;* Bracken Books, London, 1986.

Maeyama, Y., "On the astronomical data of ancient China, a numerical analysis," *Archives Intern. d'Histoire des Sciences,* 25 (1975) and 26 (1976).

Maeyama, Y., "The oldest catalogue of China, *Shi Shi Xing Jing*," Prismata-Naturwissensch. Studien, Steiner Verlag, Wiesbaden, 1977, p.211-245.

Maeyama, Y., "The development of the 28 lunar mansions in China," Int. Conf. on History of Science in China; Cambridge, 1990.

Maspero, H., "Legendes des mythologies dans le Chou King," *Journal Asiatique*, 204 (1924), pp.1-100.

Maspero, H., "L'astronomie chinois avant les Han," *Toung Pao*, 26 (1929).

Maspero, H., *Le Taoisme et Les Religions Chinoises;* Ed. Gallimard, Paris, 1953

Mathieu, R., *Anthologie des Mythes et Legendes de la Chine Ancienne;* Edition Gallimard, Paris, 1989.

Moran, H. A., and Kelley, D. H., *The Alphabet and the Ancient Calendar Signs;* Daily Press, Palo Alto, USA, 1969.

Münke, W., *Die Klassische Chinesische Mythologie;* Verlag Klett, Stuttgart, 1976.

Nakayama, Shigeru, *History of Japanese Astronomy with Chinese Background and Western Impact;* Harvard Univ. Press, Cambridge Mass., 1969.

Needham, J. et al.; *Science and Civilisation in China,* Vol. 1-3; Univ. Press, Cambridge, U.K., 1959.

Needham, Joseph; Wang Ling; Price, Derek J. de Solla. *Heavenly Clockwork*; Cambridge Univ. Press, 1960.

Needham, J., "Time and Knowledge in China and the West," *The Voices of Time*; Ed. J. T. Fraser, Braziller New York, 1966, pp. 92-135.

Ovenden, M. W., "The origin of the constellations," *Philosophical Journal*, 3 (1966), pp.1-18.

Pankenier, David W., "Reflections of the lunar aspect on Western Chou chronology," *T'oung Pao*, 78 (1992), 33-76.

Papke, W., *Die Sterne von Babylon;* Gustav Lübbe Verlag, Bergisch Gladbach, 1989.

Philippi, D., *Kojiki*; Univ. of Tokyo Press, 1968.

Pingree, D., "History of mathematical astronomy in India," *Dictionary of Scientific Biography,"* vol.15; New York, 1978.

Roy, A. E., "The origin of the constellations," *Vistas in Astronomy*, 27 (1984), pp.171-197.

Rufus, W. C., "A political star chart of the twelfth Century," *Journal of the Royal Society of Canada*, 39 (1945), p.33.

Rufus, W. C. and Hsing-chih Tien; *The Soochow Astronomical Chart;* Ann Arbor, Univ. Michigan Press, 1945.

Saussure, L. de, "La Serie Septenaire Cosmologique et Planétaire," *Journal Asiatique*, 204 (1924), pp.333-382.

Saussure, L. de, "Les Origines de l' Astronomie Chinoise (reprint of papers in T' oung Pao)," Librairie Orientale et Américaine, Paris 1925.

Schäfer, E. H., *Pacing the Void: Tang Approaches to the Stars;* Univ. of California Press, Berkeley, 1977.

Schipper, K. M. and Wang Hsiu-huei, "Time, Science and Society in China and the West," *Study of Time*, 5 (1986), pp.185-205.

Schipper, K. M., "The Taoist body," *History of Religions*, 17 (1978), pp.355-386.

Schipper, K. M., *Le Corps Taoïste*, Ed. Librairie Artheme Fayard, Paris, 1982.

Schlegel, G., *Uranographie Chinoise;* Brill, Leiden, 1875.

Shcheglov, V. P., *Star Atlas of Jan Hevelius;* Uzbek Academy of Science, Tashkent, 3rd edition, 1978.

Shen Sinyan, "Acoustics of ancient Chinese bells," *Scientific American*, 256 (1987).

Shigeru, N. and Sivin, N. *Explorations of Ancient Traditions (Needham festival)*, MIT Press, Cambridge Mass., 1973.

Sivin, N., "The Chinese conception of Time," *Earlham Review*, Vol. I, 1966, pp. 82-91.

Sivin, N., "Cosmos and computation in early Chinese mathematical astronomy," *Toung Pao,* 55 (1969).

Sivin, N., "State, cosmos, and body in the last three centuries B.C.," *Harvard Journal of Asiatic Studies,* 55 (1995), 5-37.

Sivin, N., *Science in Ancient China: Researches and Reflection;* Brookfield, Vt.: Variorum, 1995.

Soothill, W. E., *The Hall of Light: a Study of Early Chinese Kingship;* London, 1951. (Contains a translation of the *Xia Xiao Zheng.*)

Stanley, S., *The New Grove Dictionary of Music and Musicians, No.4, Chinese Music;* Mac Millan Publishers Ltd., London, 1980.

Stephenson, F.R., "Stargazers of the Orient," *New Scientist,* 1 (1993).

Stephenson, F.R., "Chinese and Korean star maps and catalogs," in J.B. Harley and David Woodward (eds.): *The History of Cartography.* Volume 2, Book 2: *Cartography in the Traditional East and Southeast Asian Societies.* Chicago: Univ. of Chicago Press, 1994, 511-578.

Teboul, M., *Les Premieres Théories Planétaires Chinoise.* (Mémoires de L'Institut des Hautes Études Chinoise, Volume XXI.) Collége de France, 1983.

Toomer, G. J., Ptolemy's *Almagest* (in English), Editor Duckworth, London, 1984.

Tsuchihashi, I. et al; "Catalogue d'etoiles fixes observées en Chine sous l'empereur K'ien Lung (1757 AD)," *Annales de l'Observatoire Astronomique de Zo-Se, Shanghai, Chine,* VII, 1911.

Veith, I.; *Classical Internal Medicine of the Yellow Emperor;* Berkeley, 1966.

Waerden, B. L. v.d. *Science Awakening,* Vol.II; Oxford Univ. Press, 1974.

Waley, A., *The Book of Songs, Sji Tsjing;* G. Allen and Unwin, London, 1954.

Waltham, C.; *Shujing, the Book of History;* Ed. Regnery and Co., Chicago, 1971.

Xi Zezong, "Chinese studies in the History of astronomy, 1949-1979," *Isis,* 72 (1981): 456-70.

Yabuuchi Kiyoshi, "Researches on the *Chiu-chih Li* — Indian astronomy under the T'ang Dynasty," *Acta Asiatica,* 36 (1979): 7-48.

Index of Star Names

Chinese star names in this index are arranged alphabetically. Each item gives in sequence: (i) star name in *pinyin* spelling; (ii) star name in Chinese characters; (iii) the Appendix in which it appears and the number in the Appendix; and (iv) Western constellation where it was located, indicated on the star maps as well.

General Index

SINICA LEIDENSIA

Edited by

W.L.IDEMA

in cooperation with

P.K.BOL · D.R.KNECHTGES · E.S.RAWSKI
E.ZÜRCHER · H.T.ZURNDORFER

11 E. Zürcher. *The Buddhist Conquest of China.* The Spread and Adaptation of Buddhism in Early Medieval China. 1. Text. 2. Notes, Bibliography, Indexes. Reprint of the 1st (1959) ed., with additions and corrections. 1972. ISBN 90 04 03478 1

12/2 W. R. B. Acker (tr.). *Some T'ang and Pre-T'ang Texts on Chinese Painting.* With Annotations. Vol. II. Chang Yen-Yüan. *Li tai ming hua chi,* Chapters IV-X. 1. Translation and Annotations. 2. Chinese text. 1974. ISBN 90 04 03938 4

13 W. L. Idema. *Chinese Vernacular Fiction.* The Formative Period. 1974. ISBN 90 04 03974 0

15 W. L. Idema (ed.). *Leyden Studies in Sinology.* Papers Presented at the Conference Held in Celebration of the Fiftieth Anniversary of the Sinological Institute of Leyden University, December 8-12, 1980. 1981. ISBN 90 04 06529 6

16 W. L. Idema. *The Dramatic Œuvre of Chu Yu-Tun (1379-1439).* 1985. ISBN 90 04 07291 8

17 A. F. P. Hulsewé. *Remnants of Ch'in Law.* An Annotated Translation of the Ch'in Legal and Administrative Rules of the 3rd Century BC Discovered in Yün-meng Prefecture, Hu-pei Province, in 1975. 1985. ISBN 90 04 07103 2

18 Ph. de Heer. *The Care-Taker Emperor.* Aspects of the Imperial Institution in Fifteenth-Century China As Reflected in the Political History of the Reign of Chu Chi'i-yü. 1986. ISBN 90 04 07898 3

19 N. Standaert. *Yung Tingyun, Confucian and Christian in Late Ming China.* His Life and Thought. 1987. ISBN 90 04 08127 5

20 H. T. Zurndorfer. *Change and Continuity in Chinese Local History.* The Development of Hui-Chou Prefecture 800 to 1800. 1989. ISBN 90 04 08842 3

21 B. J. Mansvelt Beck. *The Treatises of Later Han.* Their Author, Sources, Contents and Place in Chinese Historiography. 1990. ISBN 90 04 08895 4

22 E. B. Vermeer (ed.). *Development and Decline of Fukien Province in the 17th and 18th Centuries.* 1990. ISBN 90 04 09171 8
23 K. Ruitenbeek. *Carpentry and Building in Late Imperial China.* A Study of the Fifteenth-Century Carpenter's Manual *Lu Ban jing.* 1993. ISBN 90 04 09258 7
24 W. L. Idema and E. Zürcher (eds.). *Thought and Law in Qin and Han China.* Studies Dedicated to Anthony Hulsewé on the Occasion of His Eightieth Birthday. 1990. ISBN 90 04 09269 2
25 B. J. ter Haar. *The White Lotus Teachings in Chinese Religious History.* 1992. ISBN 90 04 09414 8
27 T. Yoshida. *Salt Production Techniques in Ancient China.* The *Aobo Tu.* Translated and revised by H. U. Vogel. 1993. ISBN 90 04 09657 4
28 Huang Chun-chieh and E. Zürcher (eds.). *Norms and the State in China.* 1993. ISBN 90 04 09665 5
29 L. Blussé and H. T. Zurndorfer (eds.). *Conflict and Accommodation in Early Modern East Asia.* Essays in Honour of Erik Zürcher. 1993. ISBN 90 04 09775 9
30 Ming-Wood Liu. *Madhyamaka Thought in China.* 1994. ISBN 90 04 09984 0
31 L. P. Edwards. *Men & Women in Qing China.* Gender in *The Red Chamber Dream.* 1994. ISBN 90 04 10123 3
32 H. Eifring. *Clause Combination in Chinese.* 1995. ISBN 90 04 10146 2
33 Huang Chun-chieh and E. Zürcher (eds.). *Time and Space in Chinese Culture.* 1995. ISBN 90 04 10287 6
34 E. L. Farmer. *Zhu Yuanzhang and Early Ming Legislation.* The Reordering of Chinese Society following the Era of Mongol Rule. 1995. ISBN 90 04 10391 0
35 Y. Ariel. *K'ung-ts'ung-tzu.* A Study and Translation of Chapters 15-23 with a Reconstruction of the *Hsiao Erh-ya* Dictionary. 1996. ISBN 90 04 09992 1
36 L. S. K. Kwong. *T'an Ssu-t'ung, 1865-1898.* Life and Thought of a Reformer. 1996. ISBN 90 04 10471 2
37 A. Hansson. *Chinese Outcasts.* Discrimination and Emancipation in Late Imperial China. 1996. ISBN 90 04 10596 4
38 Sun Xiaochun and J. Kistemaker. *The Chinese Sky during the Han.* Constellating Stars and Society. 1997. ISBN 90 04 10737 1

ISSN 0169-9563

www.ingramcontent.com/pod-product-compliance
Ingram Content Group UK Ltd.
Pitfield, Milton Keynes, MK11 3LW, UK
UKHW021409091026
15624UKWH00009B/74

9 789004 107373